Physical Science

An Introduction To Our World

By

Sheila Chustek
Adjunct Lecturer
Queens College

KENDALL/HUNT PUBLISHING COMPANY
4050 Westmark Drive   Dubuque, Iowa 52002

PREFACE

Physical Science: An Introduction to Our World is intended as a semester course for students in the Queens College, Adult Collegiate Education program known as the ACE program. It is designed as a companion to any text dealing with a conceptual approach to physics. It covers such topics as matter, mechanics, light, and heat. Only a reading ability is required to develop an understanding of each topic. In some instances, basic skills in mathematics are introduced so as to bring clarity to the principles of physics. The design of the book is such that the student will find the material and approach to be user friendly. In addition, at the conclusion of each chapter there is a set of questions reviewing the material, followed by some humorous math questions.

This book is a labor of love and came into being at the request of Bob Weller, the Director of the ACE program and the Assistant to the Provost of Queens College. I will always be grateful to him for urging me to act on the idea to write this book. I hope the pleasure I derived in organizing and putting my thoughts on paper will be equal to that which my students will acquire by using this text. I also want to acknowledge the support and encouragement of my family, all of whom provided various forms of input.

Sheila Chustek

TABLE OF CONTENTS

PAGE

Chapter I

INTRODUCTION

Physics is the science of the natural world. In addition to the language of mathematics, the physicist needs a measuring device and a timing device.

CONCEPT OF LENGTH

As originally proposed in 1670 by a French vicar named Gabriel Mouton, the meter was supposed to be one ten millionth of the distance between the North Pole and the Equator. This distance was translated into scratches on a special platinum-iridium bar when the bar temperature was 0°C. This is known as the standard meter and is kept at the International Bureau of Weights and Measures at Sevres, France, just outside of Paris, France. The Bureau of Standards in Washington, D.C. houses an accurate copy of this standard meter. However, when scientists began measuring things as small as atoms, scratches on bars no longer offered the required accuracy. The meter was internationally redefined in 1967 as 1,650,763.73 wavelengths of orange-red light emitted by the glowing krypton-86 gas. Now with the advances in laser technology, scientists have redefined the meter in terms of the speed of light. Thus, in 1983, the world's scientists accepted a new definition of the meter as the distance over which light will travel in one 299,792,458th of a second.

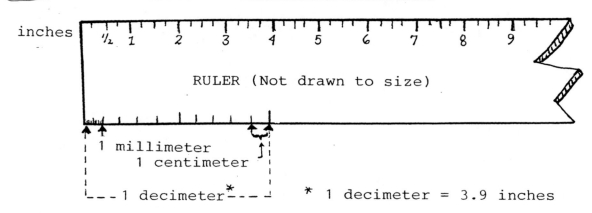

RULER (Not drawn to size)

1 millimeter
1 centimeter
1 decimeter*   * 1 decimeter = 3.9 inches

fig. 1

A meter stick is slightly longer than a yard stick since a meter is equal to 39.37 inches and a yard stick is equal to 36 inches. Some common metric measurements are as follows:

                ten millimeters = one centimeter
                ten centimeters = one decimeter
                ten decimeters = one meter
                ten meters   = one decameter
                ten decameters = one hectometer
                ten hectometers = one kilometer
                ten kilometers = one myriameter

The names are formed by adding LATIN prefixes such as:

    "deci" meaning one tenth (.1)
    "centi" meaning one hundredth (.01)
    "milli" meaning one thousandth (.001)
    "micro" meaning one millionth (.000001)

Higher amounts are formed by multiplication of multiples of ten and using GREEK prefixes as follows:

    "mega" meaning 1,000,000 (10^6)
    "myria" meaning   10,000 (10^4)
    "kilo" meaning   1,000 (10^3)
    "hecto" meaning   100 (10^2)
    "deka" meaning   10 (10^1)

Some prefix applications are:

    A microsecond = one millionth of a second
    A milligram   = one thousandth of a gram
    A centimeter = one hundredth of a meter
    A kilogram   = one thousand grams
    A megahertz   = a million hertz

MEMORY TRICKS:

-The meter is about the distance from the doorknob to the floor.
-A golf club is about one meter in length.
-As the dollar is divided into 100 equal parts called cents,

the meter is divided into 100 equal parts called
centimeters.
-As the dollar is divided into ten equal parts called dimes,
the meter is divided into ten equal parts called
decimeters.

CONCEPT OF TIME

The standard of time interval used in scientific work is
based on the mean solar day. The solar day is the time
elapsed between two successive noons. Since solar days vary
slightly at different times of the year, the mean solar day
is an average value taken over the year. The second, which
is the standard unit of time, is 1/86,400th of the mean
solar day. If each day has 24 hours and each hour has 60
minutes, then each day has 1,440 minutes.

$$24 \; \frac{\text{hours}}{\text{day}} \quad X \quad 60 \; \frac{\text{minutes}}{\text{hour}} \quad = \quad 1,440 \; \frac{\text{minutes}}{\text{day}}$$

Since each day has 1,440 minutes and each minute has 60
seconds, then each day has 86,400 seconds.

$$1,440 \; \frac{\text{minutes}}{\text{day}} \quad X \quad 60 \; \frac{\text{seconds}}{\text{minute}} \quad = \quad 86,400 \; \frac{\text{seconds}}{\text{day}}$$

Thus a second is 1/86,400th of a day. Today, however, the
second is defined as the duration of 9,192,631,770 cycles of
the radiation associated with a specified transition of the
cesium-133 atom.

CONCEPT OF MASS

The standard for the unit of mass, the kilogram, is a cube
of platinum-iridium alloy kept by the International Bureau
of Weights and Measures at Sevres, France. The instrument
used in comparing mass is the delicate balance, which is an
equal-arm balance. Two masses are compared and if a balance
is met, they are equal. Two accurate copies of the standard

kilogram are in the custody of the National Bureau of Standards in Washington, D.C. and this serves as the mass standard for the United States. This is the only base unit still defined by an artifact.

SYSTEM OF UNITS

The meter, as the unit of length, the kilogram, as the unit of mass, and the second, as the unit of time make up the MKS system of units. This is part of the International System of units known as S.I. for Systeme International. This is a modernized version of the metric system established by international agreement. It provides a framework for all measurements in science, industry, and commerce. The system is built upon a foundation of base units from which all other S.I. units are derived. The base units are called fundamental units and the others are referred to as derived units. There are two other systems of base units which are not international systems but are used in certain countries. One of these systems is the CGS system. "C" is for centimeter as a unit of length, "G" is for gram as a unit of mass, and "S" is for second as a unit of time. The other system is the FPS system. "F" is for foot as the unit of length, "P" is for pound as the unit of weight not mass, and "S" is for second as the unit of time.

Once you define the meter, then other measurements can be determined based on the meter. If the original length of the meter were to be changed then all measurements which are multiples or portions of the meter would also change. Another way to explain this is to examine Euclid's Famous Fifth Postulate, known as Euclid's Parallel Postulate. This postulate states that "through a point P outside a given line l there is one and only one line parallel to the given line l." Parallel lines are defined as lines that lie in the same plane and never meet no matter how far they are extended in either direction. The Parallel Postulate is assumed to be true as are all postulates and they are all part of the foundation of Euclidean Geometry. Remember, postulates are statements which are accepted as true without proof. Many mathematicians did not feel that the Parallel

Postulate was self-evident and believed it could be proved or disproved. A Russian mathematician named Lobachewsky, who lived in the early 1800's, believed that if the postulate could not be proved, it could be changed. He assumed that "through a point P outside a given line l there exist an infinite number of lines parallel to the given l". Then with this new postulate replacing Euclid's Parallel Postulate and with all of Euclid's other postulates, he constructed a different but consistent geometry with no contradictions. At the same time, but independently, a Hungarian mathematician, named Johann Bolyai, worked on the same concept of this non-Euclidean geometry. He developed his own but similar geometry to Lobachewsky. Again, at about the same time, a German mathematician named Reimann, constructed another geometry in which he also changed Euclid's Parallel Postulate but in a different way. Reimann's postulate stated that "through a point P outside a given line l there is no line parallel to the given line l". These new geometries are referred to as non-Euclidean geometries. It should be understood that once a postulate, a statement accepted as true without proof, is changed, the entire geometry is not the same. A good example to illustrate this point is the theorem about the sum of the angles of a triangle. A theorem is a statement in geometry which is proven true based on assumptions known as the postulates. In Euclid's geometry the sum of the angles of a triangle is equal to 180°. In the geometry resulting from Bolyai and Lobachewsky's new postulate, the sum of the angles of a triangle is less than 180°. In Reimann's geometry, the sum of the angles of a triangle is greater than 180°. All three geometries are equally good, but they are not all equally convenient in various applications. For ordinary everyday needs, the Euclidean geometry is the best. Einstein found the Reimannian geometry best fit his needs. Prior to these investigations, there was an Italian mathematician, named Saccheri, who in the late 1700's, set out to prove that the Parallel Postulate could be proven true from the previous postulates. However, he proved just the opposite. Saccheri found that the Parallel Postulate is independent of the other postulates and, therefore, cannot be deduced from them. In other words it cannot be proven and must remain as a postulate.

Over the years, physicists may have changed the method of arriving at the length called the meter but the actual length has never changed. To do so would change physics as we know it today.

FORCE - VECTORS

Force is a concept which requires knowledge of the ideas of length, time and mass. Force is needed to support an object, to push an object or to pull an object. The force with which the Earth pulls an object down is called the weight of the object. The unit of force in the MKS system is called a newton. The unit of force in the CGS system is called a dyne. A more thorough explanation of these terms will be given in the Chapter III which deals with Newton's laws of motion.

One way of describing a force is by its size and direction. The size is the amount of the force and is referred to as the magnitude of the force. The magnitude of a force is measured in units called newtons and is usually abbreviated with a capital "N". For example, if you push a box with a force of ten newtons, the magnitude of the force is 10N. The direction is also important. Any quantity that has both magnitude and direction is called a vector quantity. Forces consist of both magnitude and direction and are therefore represented by vectors. A vector is drawn as an arrow where the arrow head shows the direction and the arrow length represents the magnitude or size of the force. For example, in order to draw a vector showing a force of 10N pushing east where 1 cm = 2N, a 5 cm line should be drawn with the arrow pointing to the right.

VECTOR ADDITION

Vectors represent quantities that have magnitude and direction. Some quantities have no direction such as time, temperature, energy and volume and are therefore not vector quantities. They are called scalar quantities. A scalar quantity requires the amount which is the magnitude and the units to represent this amount. Of course, a vector

requires the direction as well. In addition to using a
vector to represent a force, a vector can also be used to
show displacement because it has magnitude and direction.
Displacement is the distance and the direction from a given
starting point to a finishing point. For example, if you
walk two miles north from your home, your displacement is
two miles north. If you walked three miles north and then
four miles east, you walked a total of seven miles.
However, your displacement is not seven miles. Your
displacement is the shortest distance from your starting
point to the point where you end no matter what path is
taken. In this example, it is 5 miles at an angle 53° east
of north and is called the vector sum of the other two
vector quantities. It is calculated by using the
Pythagorean Formula and trigonometric functions (fig. 2)

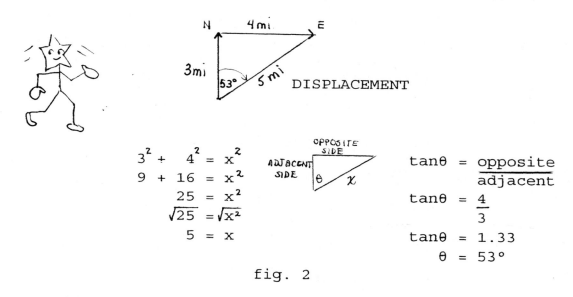

$$3^2 + 4^2 = x^2$$
$$9 + 16 = x^2$$
$$25 = x^2$$
$$\sqrt{25} = \sqrt{x^2}$$
$$5 = x$$

$$\tan\theta = \frac{opposite}{adjacent}$$
$$\tan\theta = \frac{4}{3}$$
$$\tan\theta = 1.33$$
$$\theta = 53°$$

fig. 2

In the figure above (fig. 2), vectors are used to represent
the walking path and the displacement. The displacement
vector represents the sum of the other two vectors.
However, when vectors are used to represent forces, the sum
of the forces is represented by a vector called the
resultant. The combination of two or more forces acting on
an object is called the resultant of the forces.
Consequently, there are two methods for adding vectors. The
first method is used to compute displacement as shown above
and is also used to find the resultant if "a" and "b"
represent successive forces. It is called the polygon
method of adding vectors. Place "b" such that its tail is

at the head of "a" and draw the vector, the resultant, from
the tail of "a" to the head of "b" (fig. 3). Secondly,
there is the parallelogram method of adding vectors and is
used when two forces "a" and "b" are acting simultaneously
on an object. Place "b" such that its tail is at the tail
of "a" and complete the parallelogram, Now draw the
resultant along the diagonal (fig. 3).

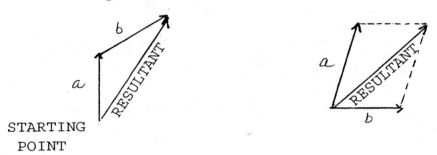

STARTING
  POINT

fig. 3

For example, suppose there is a 90° angle between two forces
of 6N an 8N acting simultaneously on a point mass. The
resultant would be along the diagonal of the completed
parallelogram (fig. 4). The angle between the resultant and
either of the two forces could be measured using a protractor
or could be computed with the use of trigonometry. The
magnitude of the resultant could also be computed by laws
of trigonometry. Sometimes three or more forces act at
different angles on the same object at the same time. Can
you think of a way to find the resultant, if the concurrent
forces are acting the same plane? not in the same plane?

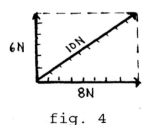

fig. 4

Vector addition is simplified when the forces are in the
same or opposite directions. Suppose one person pulls on a
rope attached to an object with a force of 4N eastward.
Another person pulls on the same object in the same
direction with a force of 10N. The resultant force is the
same as if one person pulled on the object in that same
direction with a force of 14N (fig. 5a). Suppose the same

people pulled with the same amount of force in opposite directions. The resultant is a force of 6N eastward (fig. 5b). Now suppose the same people are pulling in opposite directions but with forces of 10N each. The resultant force would be zero N (fig. 5c).

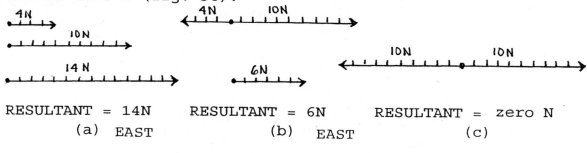

RESULTANT = 14N   RESULTANT = 6N   RESULTANT = zero N
    (a) EAST   (b) EAST   (c)

fig. 5

 AH HA!

EXAMPLES FOR THOUGHT I

1. Use an encyclopedia to research the history of the origin of the standard meter.

2. Which vector below best represents the resultant of the two vectors shown?

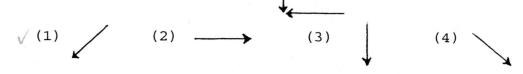

  √ (1)   (2)   (3)   (4)

3. If a baseball player runs 90' from home to first base then over-runs 15' and returns to first. The distance that he has run is 120'. How many feet is his displacement? 90'

4. As the angle between two concurrent forces decreases from 180°, their resultant
    (1) decreases (2) increases (3) remains the same

5. The velocity vector shown in the diagram below is resolved into two components, these two components could best be represented by which diagram?

   (1)   (2)   (3)   (4)

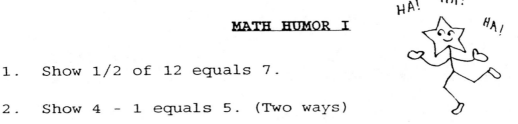

6. If two parallel forces of 20N and 10N, respectively, are in equilibrium with a force of 30N, the resultant of **the three forces** is

   (1) zero N   (2) 10N   (3) 60N   (4) 30N

7. Find the sum of 3.5m east and 5.8m south. Give both magnitude and direction of the sum.

MATH HUMOR I

1. Show 1/2 of 12 equals 7.

2. Show 4 - 1 equals 5. (Two ways)

3. A solid brick has six surfaces. If the brick is cut in half, how many surfaces would each half have?

Chapter II

LINEAR MOTION

Which is larger 1/7 or 1/8?

MEMORY TRICKS

Set up the fractions as follows:

Multiply the first numerator of
1 by the second denominator of 8
and put the answer under the 1/7.
Now multiply the second numerator
by the first denominater and put
the answer under the 1/8. The

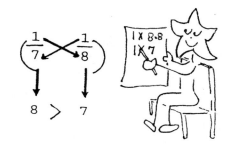

fraction above the larger of the two answers is the larger
fraction.   The reason this works is that you are comparing the
numerators without writing the common denominator which is 56.

Hold a dollar bill between another
person's fingers so that they do not
touch the bill as illustrated in
fig. 1 at the right and then drop the
bill.   The other person would have a
difficult time catching the dollar bill
because the time needed for half the
bill to fall past the person's hand is
1/8th of a second and the person's
reflex time is 1/7th of a second.   The
example above with the fractions showed
that 1/7 is greater than 1/8 which means
that the response time is greater than
the time required for the top half of the
bill to pass below the closing fingers.

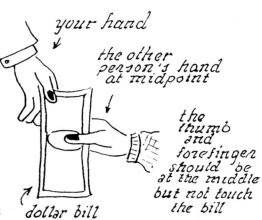

your hand

the other person's hand at midpoint

the thumb and forefinger should be at the middle but not touch the bill

dollar bill

fig. 1

The dollar bill was used to illustrate a common notion that when an item is not held or supported it will fall downwards, not sideways and not upwards. Everyone knows it will fall downwards but the question is why and at what rate? This chapter will examine these questions.

ARISTOTLE AND MOTION

According to Aristotle, there exists NATURAL MOTION and VIOLENT MOTION. In natural motion everything has a place. If an object is not in its natural place, it would move until it got there. For example, the dollar bill in the previous example did not belong in a floating position and would, therefore, fall to a resting position on a surface where it belonged. On the other hand, violent motion occurs when an object is pulled or pushed. The cause of the push or pull is not always obvious.

With these notions, it appeared that Aristotle believed all objects belonged at rest. Objects in their proper place would not be in motion. Consequently, the Earth must not be moving. This concept gave us the geocentric theory of the universe which states that the Earth is the center of the Universe and the Sun revolves around it.

Another astronomer named Copernicus found from observations that the Sun is the center of the universe and that the Earth revolves around the Sun. This is the heliocenteric theory of the Universe. Because he was reluctant to contradict Aristotle's views, Copernicus never published his findings. They were later published in 1543.

It was another scientist named Galileo who backed the Copernican view that the Sun is the center of the Universe. After much research he was able to establish that Aristotle's ideas had some weaknesses. For example, he found fault with Aristotle's concept that an object could only move if it were pushed or pulled. Through experimentation, Galileo was able to show that in the absence of any interferring force an object would move forever and ever in a straight line. No push or pull was required to keep the object moving once it got started.

Galileo used two inclined planes and a flat surface as shown in the diagram at the right. In each case, the initial roll of the ball down the inclined plane was the same. In the first case, (fig. 2), where the inclined plane at the right was the steepest, he found the ball rolled to about the same height as the initial height. In the second case, (fig. 3), the slope of the inclined plane at the right was less and the ball also rolled to about the same

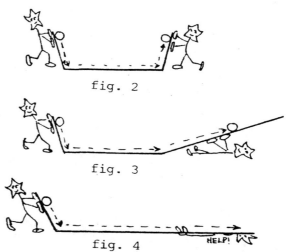

fig. 2

fig. 3

fig. 4

height as the initial height but rolled further to acheive this height. He concluded that the ball must have lost its speed more slowly. Finally, in the third case, (fig. 4), there was no incline at the end and he found the ball continued to roll indefinitely. From this, he concluded that the ball was not losing any speed and would continue to roll forever and ever. In the absence of an external force, the motion of the ball would continue in a straight line. This gave rise to the concept that Galileo referred to as **inertia**.

DESCRIPTION OF MOTION

Three ideas are needed to describe motion. They are **speed, velocity** and **acceleration**.

$$\text{SPEED} = \frac{\text{DISTANCE}}{\text{TIME}}$$

The speed at any one instant is called <u>instantaneous speed</u>. It is the speed registered on the speedometer of a car.

    An example is: 80 miles per hour (80 mph)

    or in the S.I. system: 80 kilometers per hour (80 km/hr)

The speed is the rate in the formula **rate X time = distance** which we learned in grade school. Dividing both sides of this equation by **time**, we get:

$$\textbf{RATE } = \frac{\textbf{DISTANCE}}{\textbf{TIME}}$$

This matches the formula above for speed.

When we specify speed with direction and not just speed, we are referring to the **velocity**. Velocity is speed and direction.

For example: **VELOCITY** = 80 km/hr to the north
whereas   **SPEED**   = 80 km/hr

If we speak of a constant velocity, we are implying a constant speed (no change in speed) and a constant direction (no change in direction). If an object has a constant speed and does not have a constant direction then it does not have a constant velocity. For example, a car may be traveling at a constant speed in a circle. The direction is not constant because the car's direction is changing in order to stay in the circular path. There may also be a case when the direction is constant, but the speed is not constant which again means that the veloicty is not constant. An example is a car slowing down or speeding up while moving on a straight road. The direction is not changing, but the speed is changing. Another case when the velocity is not constant is when both the speed and the direction change such as on a roller coaster where the car goes slower and faster while going uphill and downhill. In summary, a constant velocity requires a constant speed and a constant direction. If there is a change in the speed, or a change in the direction, or a change in both the speed and the direction then the velocity is not constant. When we make any of these changes we are changing the velocity and the rate at which any of these changes occur is called acceleration.

$$\textbf{ACCELERATION } = \frac{\textbf{CHANGE IN VELOCITY}}{\textbf{TIME}}$$

Acceleration is the change in velocity with respect to time.

ACCELERATION

CASE I	CASE II	CASE III
CHANGE IN SPEED NOT DIRECTION	**CHANGE IN DIRECTION NOT SPEED**	**CHANGE IN SPEED AND DIRECTION**

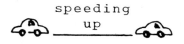

40 km/hr to 80 km/hr	40 km/hr in a circle	slower and faster up and down

If you ride on a bus you might experience acceleration when the bus driver speeds up or slows down or when the driver turns a corner. Acceleration is the rate at which the velocity changes. When slowing down, acceleration may be referred to as deceleration or negative acceleration.

We now need to know what units are used when referring to acceleration. To do this, suppose we are in a car driving at 40km/hr (40 kilometers per hour). In the next second we increase our speed to 45km/hr. This means that in one second we increased our speed by an amount of 5km/hr. Suppose we continue to increase or change our speed by 5km/hr each second. We call this acceleration and can say our acceleration is equal to five kilometers per hour per second (5 km/hr/s). Notice that there are two units for time because every second the velocity is changing by an amount of 5km/hr.

	change: 5 km/hr time: 1 sec.		change: 5 km/hr time: 1 sec.		acceleration:
40 km/hr		45 km/hr.		50 km/hr	5 km/hr/sec.

fig. 6

ACCELERATION OF A FALLING OBJECT

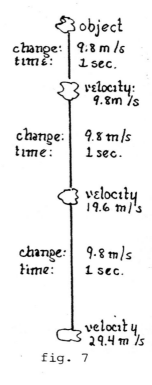

fig. 7

Galileo found that, regardless of weight or size, all objects, in the absence of air resistence, fall with a constant acceleration. Notice that we are referring to a constant acceleration and not a constant velocity. The acceleration is not changing when it is a constant acceleration. It is, the velocity that continually changes, but the amount of change in a given time interval is the same. The constant acceleration for all falling objects, in the absence of air resistance, is equal to 9.8 meters per second per second (9.8m/s/s or 9.8m/s²). This means that for every second an object is in free fall, it increases its' velocity by 9.8m/s. Each second the same amount of 9.8m/s is added to the velocity. That is, the change in velocity is 9.8m/s for every second that the object is in free fall.

A lower case letter "g" is used to represent this acceleration since it is due to gravity. Acceleration is not always due to gravity. For example, the acceleration arrived at by pressing the accelerator pedal in a car is not due to gravity and is represented by an "a". However, when it is due to gravity, we use the "g" which is equal to 9.8meters/second squared. The value of "g" is slightly different for various locations on or above the earth's surface. The value of "g" is not just slightly different but is totally different on other planets.

For example: on Mars "g" is   3.3m/s²
                 on Jupiter "g" is 25.6m/s²

ACQUIRED VELOCITY FOR AN OBJECT IN FREE FALL

(Remember the increase in velocity each second is 9.8m/s.)

Object at rest: velocity =   0m/s
At the end of the 1st sec.:   v = 9.8m/s   (9.8m/s² X 1s)
At the end of the 2nd sec.:   v = 19.6m/s   (9.8m/s² X 2s)
At the end of the 3rd sec.:   v = 29.4m/s   (9.8m/s² X 3s)

Notice that each second the velocity increases by 9.8m/s. This is the acceleration. In other words, at the end of each second the object falls 9.8 m/s faster than it fell in the second, before. A falling object has a speed of 9.8m/s after falling for 1 second. It has a speed of 9.8m/s plus 9.8m/s after falling for 2 seconds, etc. An object falling from a great height accelerates to a great speed and strikes the Earth with a great force. That is why an object that falls only a short distance may not break, but it may break when it falls a long distance. Looking again at the table above, the last column shows that instead of adding 9.8m/s for each second, the acquired velocity is equal to the rate of acceleration (9.8m/s^2) times the number of seconds during which the acceleration acts. An object falling for five seconds has an acquired velocity of 9.8m/s^2 X 5s or 49m/s at the end of the fifth second. Thus, the

$$\text{acquired velocity = acceleration X time}$$
$$\text{or} \quad v = at$$
$$\text{or} \quad v = gt \quad (g = \text{acc. due to gravity})$$

To put an object into free fall, you do not have to just drop it. You can also throw it up, throw it down (as opposed to dropping it), or throw it sideways (horizontally). Once it leaves your hand, it is in free fall and accelerates downward at the rate of acceleration due to gravity which again is 9.8m/s^2.

THROWING AN OBJECT STRAIGHT UP

Suppose you throw an object straight up into the air. It is moving up, but gravity is acting on it in the opposite direction. If the object left your hand going up at 25m/s, then after 1 second it is going at 25m/s - 9.8m/s = 15.2m/s upwards. It loses another 9.8m/s in the next second, etc.. Eventually it will reach a plateau and start coming down at the rate of acceleration of 9.8m/s^2.

DISTANCE AND ACCELERATION

If you were driving at 60mph for 2 hours, then you drove 120 miles. (60mph X 2hr = 120mi). In general, if the velocity is constant, then the distance an object moves for a certain time interval is given by the formula:

$$\textbf{distance = velocity X time}$$

or $$\textbf{d = vt}$$

Note: This is the same formula for distance (distance = rate X time; d=rt) referred to in the beginning of this chapter and it works very well for a constant velocity.

When an object is accelerating, such as, an object in free fall, the formula above will not work. The formula needed now is:

$$\textbf{distance traveled} = \textbf{½(acc.)(time)(time)}$$

or $$\textbf{d = ½at}^2$$

or $$\textbf{d = ½gt}^2 \text{ if the acceleration is due}$$
$$\text{to gravity.}$$

This distance formula is necessary because the velocity is not constant throughout the given time interval. For example, if an object is in free fall starting its fall at 0 m/s, then its velocity is 9.8 m/s only in the last instant of the one second interval. Its average velocity during the interval is the average of its beginning and final velocities in that interval.

STARTING FROM REST AND FALLING FREELY

The following values are for an object falling freely for 3 seconds. Using the formula $d = ½gt^2$ we get:

1st sec.	$d = ½(9.8m/s^2)(1s)^2 = 4.9m$;	(vel. acquired = 9.8m/s)
2nd sec.	$d = ½(9.8m/s^2)(2s)^2 = 19.6m$;	(vel. acquired = 19.6m/s)
3rd sec.	$d = ½(9.8m/s^2)(3s)^2 = 44.1m$;	(vel. acquired = 29.4m/s)

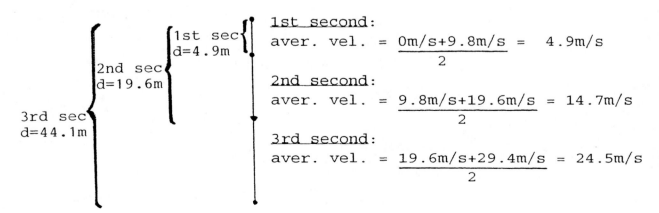

1st second:

aver. vel. $= \dfrac{0m/s + 9.8m/s}{2} = 4.9m/s$

2nd second:

aver. vel. $= \dfrac{9.8m/s + 19.6m/s}{2} = 14.7m/s$

3rd second:

aver. vel. $= \dfrac{19.6m/s + 29.4m/s}{2} = 24.5m/s$

Let's see how the distance formula works in an example.
Suppose a cat falls out of a tree, dropping 19.6m to the ground. How long is the cat in the air?

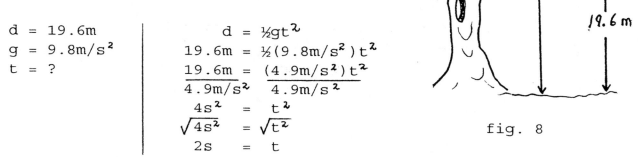

$$d = 19.6m$$
$$g = 9.8m/s^2$$
$$t = ?$$

$$d = \tfrac{1}{2}gt^2$$
$$19.6m = \tfrac{1}{2}(9.8m/s^2)t^2$$
$$\frac{19.6m}{4.9m/s^2} = \frac{(4.9m/s^2)t^2}{4.9m/s^2}$$
$$4s^2 = t^2$$
$$\sqrt{4s^2} = \sqrt{t^2}$$
$$2s = t$$

fig. 8

The cat is in the air for 2 seconds.

THROWING A BALL HORIZONTALLY

Suppose an individual throws a ball in a horizontal path to another individual. In the absence of air resistance the horizontal speed does not change. This is because gravity affects the vertical velocity but does not change the horizontal velocity. The ball's horizontal velocity remains constant but vertically it is in free fall and is accelerating downward at the rate of $9.8m/s^2$. The ball loses speed going up and gains speed coming down at the same time that it moves horizontally at a constant speed. Gravity does not affect the velocity of an object at all when the object is moving perpendicular to the pull of gravity (that is, at right angles to the vector representing the pull of gravity). If an object is thrown horizontally and not up in the air it would hit the ground at the same time as if it were dropped from rest. The object which is thrown horizontally will continue to move in that direction while accelerating downward at the same rate as an object which is just dropped.

To see how this works let's do the following example.
A baseball is thrown from the roof of a building at 25m/s in

a horizontal direction. It
strikes the ground 2.0 seconds
later.
Find: (a) How far from the base
        of the building the
        baseball lands.
      (b) The vertical distance
        the ball fell.

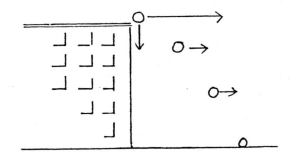

fig. 9

Hor vel. = 25m/s	Vertical: d = ½gt^2
t = 2.0s	d = ½(9.8m/s^2)(2.0s)^2
Vert. d = ?	d = 4.9m/s^2(4.0s^2)
Hor. d = ?	d = 19.6m
g = 9.8m/s^2	Horizontal: d = vt
	d = (25m/s)(2.0s)
	d = 50.0m

This tells us that the baseball landed 50.0 meters from the base
of the building while falling a vertical distance of 19.6 meters.
Notice the use of the different formulas for distance. The first
formula takes into account acceleration because the falling
object is accelerating due to gravity and the second is used
because the horizontal velocity is constant. The horizontal
velocity remains at a constant (25m/s for this example) because
the vector representing the horizontal velocity is perpendicular
to the vector representing the pull of gravity. Remember that
anytime gravity pulls on an object at right angles to the
objects' direction, it does not affect the velocity of the object
in that direction. Therefore, in this example, the horizontal
velocity of 25m/s remains constant.

THROWING A BALL DOWNWARD

Let's try an example.
    A student drops an object from the
    top of a building which is 44.1
    meters high. How long does it take
    the object to fall to the ground?

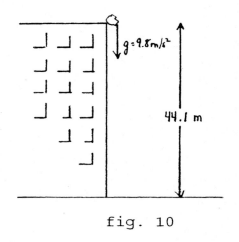

d = 44.1m	d = ½gt^2
g = 9.8m/s^2	44.1m = ½(9.8m/s^2)t^2
t = ?	44.1m = (4.9m/s^2)t^2
	$\dfrac{44.1m}{4.9m/s^2} = \dfrac{(4.9m/s^2)t^2}{4.9m/s^2}$

fig. 10

$$9s^2 = t^2$$
$$\sqrt{9s^2} = \sqrt{t^2}$$
$$3s = t$$

Thus, it took three seconds to fall 44.1m.

Another way to look at this problem is to break it down to individual seconds.

In the beginning:
        acquired vel. =   0m/s
        dist. in beginning = 0m   total dist.= 0m
  At the end: 1st sec.:
        acquired vel. = 9.8m/s
        dist. in 1st sec. = 4.9m   total dist.= 4.9m
  At the end: 2nd sec.:
        acquired vel. = 19.6m/s
        dist. in 2nd sec. = 14.7m   total dist.=19.6m
  At the end: 3rd sec.:
        acquired vel. = 29.4m/s
        dist. in 3rd sec. = 24.5m   total dist.=44.1m

This example examines the situation in which a ball is dropped as opposed to being thrown straight down. If the ball is dropped, its acceleration toward the ground is 9.8m/s^2. If instead the ball is thrown vertically downward, its acceleration is the same which is 9.8m/s^2. The acceleration is the same regardless of the initial velocity. The velocity will be different for each second of free fall, but not the acceleration. (Assume air resistance is not a factor.) In both cases the same amount is added to the velocity, but in the second case the initial velocity is not zero.

EXAMPLES FOR THOUGHT II

1. An object moves with a constant velocity of 9.8m/s. Its acceleration is in m/s^2?

$\sqrt{}$ (1) zero   (2) 4.9   (3) 9.8   (4) 32

2. Which quantity has both speed and direction?
   (1) distance   (2) speed   (3) mass $\sqrt{}$(4) velocity

3. A car starting from rest accelerates at the rate of $7m/s^2$.
   (a) How far will the car have traveled at the end of five seconds? *87.5 m*
   (b) What is the velocity in m/s that the car will acquire?
   35 m/s

4. You drop a rock from a bridge and it hits the water 3.5 seconds later. Find (a) the height of the bridge *60.025 m*
   (b) the velocity of the rock when it hits. *30.3 m/s*

5. (a) Ignoring the friction of the air, how far does an object dropped from an airplane fall in 10 seconds? *490 m*
   (b) What velocity does it acquire? *98 m/s*
   (c) How far does it fall during the last second? *93.1*

6. Give an example of a car that is accelerating while traveling at a constant speed.

7. If an object is dropped, it will accelerate downward at a rate of $9.8m/s^2$. If instead it is thrown downwards, its acceleration (in the absence of air resistance) will be
   (1) greater   (2) less   (3) the same   (4) none of these
                             $\sqrt{}$

MATH HUMOR II

1. Prove: 101 - 102 = 1 by moving one line or digit.

2. How do you make seven even?

3. The math teacher said Johnny is very smart. Who is smart in the preceding sentence?

4. Which 'that' in the following sentence should be in italics?
   "I said that that that that that man wrote should have been in italics."

5. What letter should replace the "?"?

T	H	U	Z	I	S	E	O	U	A
P		E		D		Q		?	
S	I	L	Z	A	M	S	F	E	R

Chapter III

NEWTON'S LAWS OF MOTION

To introduce the next concept, an experiment could be conducted using four black checkers and one red checker. The checkers should be stacked with the red checker on the bottom as shown in the diagram at the right (fig. 1). (Note: The checkers should have very smooth surfaces to eliminate as much friction as possible.) Now a ruler could be used to hit the red checker, forcing it to move out from under the stack. If done quickly, the four black checkers should remain stacked. In order to explain what actually took place we turn to Newton's Laws of Motion.

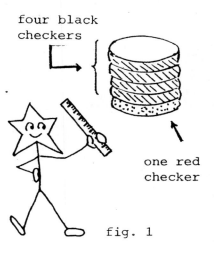

four black checkers

one red checker

fig. 1

NEWTON'S FIRST LAW OF MOTION

LAW I (THE LAW OF INERTIA):

EVERY OBJECT CONTINUES IN ITS STATE OF REST, OR OF UNIFORM MOTION IN A STRAIGHT LINE, UNLESS IT IS COMPELLED TO CHANGE THAT STATE BY FORCES IMPRESSED UPON IT. (Law of Inertia)

The concept of an object resisting change was first introduced by Galileo who called this inertia. An object does not accelerate itself. Nothing moves unless it is pushed or pulled. What happens if a wagon is stacked with newspapers and someone quickly pulls the wagon (fig. 2)? The wagon goes forward and the top newspapers fall off the back of the wagon (fig. 3). What actually happens is that the wagon is put into motion by the pull, but the newspapers are not and because of inertia they want to remain at rest. Since the wagon is pulled out from under the newspapers, the newspapers fall to the ground. They appear to slide backwards because friction sets some papers into forward motion. The same situation existed in the experiment above with the checkers. The red checker is pushed out from the bottom of the stack and the black checkers just move down to the table top. These examples illustrate the first part of the Law of Inertia which states that an object at rest will remain at rest unless an external force is imposed upon it.

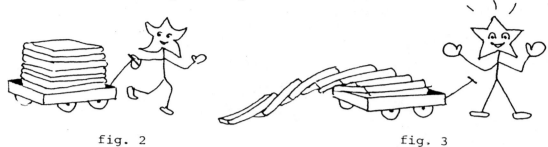

fig. 2   fig. 3

In the next scenerio, suppose the wagon with the stack of newspapers is already in motion and suddenly the wagon is forced to stop (fig. 4). The wagon is stopped but the newspapers fall forward off the front of the wagon (fig. 5). The second part of the Law of Inertia states that an object will continue to move in a straight line unless an external force is imposed upon it. The wagon is forced to stop but the newspapers continue to move forward and with no wagon underneath they fall to the ground in front. You, too, may have experienced an application of this law when standing on a moving bus which suddenly makes a short stop. You feel your body falling forward. This is because the brakes of the bus are working on the bus and not on you.

fig. 4   fig. 5

Finally, the third part of the Law of Inertia about an object moving in a straight line can be explained by visualizing yourself sitting in the front passenger seat in a car which is about to make a left turn. As the car turns left you feel yourself being thrown toward the door on the passenger side of the car. In actuality, the car was made to turn left, but you, the passenger, wanted to go straight and the passenger door got in your way (fig. 6).

fig. 6   fig. 7

A circus rider standing on a horse knows that if the rider and the horse both have the same forward speed, the rider could jump in the air, spin around and still come down on the horse's back (fig. 7). This is because the rider and the horse keep moving at the same forward speed.

Suppose an airplane is already in flight, traveling parallel to the ground and suppose at some point it drops a package. If the plane is casting a shadow on the ground, then the shadow of the package will always be within the airplane's shadow. The airplane and the package have the same forward motion and therefore, their shadows also have the same forward motion. Of course, the effects of air resistance are not being considered. Actually, the package is falling as it is moving forward and its' path is that of half of a parabola (fig. 8).

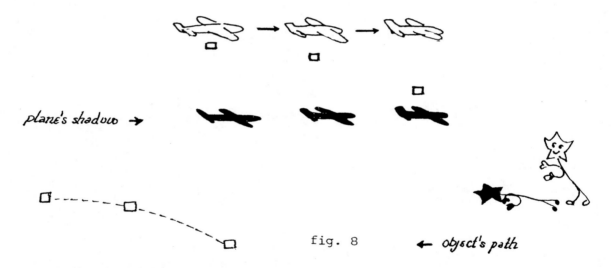

plane's shadow →

fig. 8   ← object's path

Newton saw natural motion different from the way that Aristotle saw it. Newton defined natural motion as maintaining a constant speed along a straight line and thus requiring no force. Aristotle believed in natural motion versus motion caused by some impressed force. If a planet moved in a circular path it did so according to Aristotle because it was natural for it to do so and no force was necessary. Newton believed that since the motion of the earth was not in a straight line, then some external force was necessary to keep it moving in a curved path, otherwise, the earth would move in a straight line and go off into space. In another chapter, we will discuss the laws regarding this external force as described by Newton.

NEWTON'S SECOND LAW OF MOTION

LAW II (THE LAW OF ACCELERATION):

THE ACCELERATION OF AN OBJECT IS DIRECTLY PROPORTIONAL TO THE
NET FORCE ACTING ON THE OBJECT AND INVERSELY PROPORTIONAL TO
THE MASS OF THE OBJECT AND IS IN THE DIRECTION OF THE NET FORCE.

$$acceleration = \frac{force}{mass}$$

or

$$a = \frac{f}{m}$$

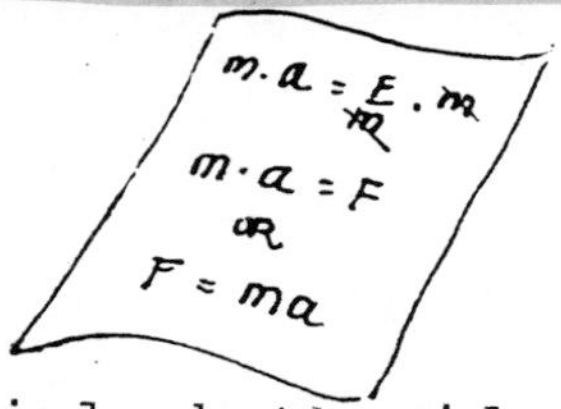

$$m \cdot a = \frac{f}{m} \cdot m$$
$$m \cdot a = f$$
or
$$F = ma$$

fig. 9

which can be written as $f = ma$ (Multiply both sides of the
equation by m.)

Force is anything that can accelerate an object which means it is
anything that can change the state of motion of an object such as
starting or stopping the motion of the object. For a given mass,
the larger the force the greater the acceleration. Twice the
force will produce twice the acceleration. If a force of 6N
acting on a 2kg object produces an acceleration of $3m/s^2$, then a
force of 12N acting on the same object will produce an
acceleration of $6m/s^2$ (fig. 10a & 10b). Mass has the opposite
effect. For a given force, twice the mass will produce half the
acceleration. For example, if a given force produces an
acceleration of $12m/s^2$ applied to a 9kg mass, it will produce an
acceleration of $6m/s^2$ applied to an 18kg mass. In other words,
if there is acceleration, the amount of acceleration depends on
the amount of force and the amount of mass.

$$A = \frac{F}{m}$$

$$a = \frac{f}{M}$$

**ACCELERATION IS DIRECTLY
PROPORTIONAL TO THE FORCE**

**ACCELERATION IS INVERSELY
PROPORTIONAL TO THE MASS**

In Chapter II, acceleration = $\frac{change\ in\ velocity}{time}$

In this chapter, acceleration = $\frac{force}{mass}$

This new formula shows another way of looking at acceleration. A
net force applied to an object causes the object to be

accelerated. If the net force acting on the object is not zero, the object will be accelerated in the direction of the net force.

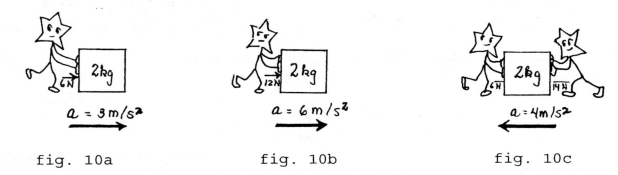

fig. 10a   fig. 10b   fig. 10c

If the force is doubled for a given mass, then the acceleration is also doubled (fig. 10a and 10b). In addition the direction of the acceleration is in the direction of the net force (fig. 10c).

If an object is in free fall the acceleration of the object is "g". The force of this downward acceleration is called the object's weight.

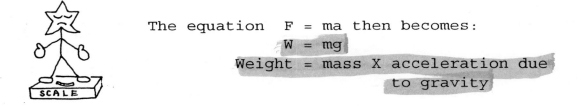

The equation F = ma then becomes:
$$W = mg$$
Weight = mass X acceleration
                    to gravity

Weight varies from location to location and is directly proportional to the acceleration due to gravity. But weight also depends on the mass, since objects may have different weights even in the same place where gravity is the same. Twice the mass will weigh twice as much at a given location. Weight depends on both mass and the acceleration due to gravity.

NEWTONS

As mentioned in Chapter I, mass is measured in kilograms, acceleration is measured in m/s^2, and force is measured in Newtons. Now for an explanation of Newtons:
$$F = mass\ X\ acceleration$$
$$F = kg\ X\ m/s^2$$

Measuring force in units of kg X m/s^2 may seem cumbersome, but they are perfectly good units and they keep the equations consistent. However, this collection of units has been given a name which is the Newton, abbreviated as "N".

$$1 \text{ Newton} = 1 \text{ kg X m/s}^2$$

One Newton is defined as that force which will give a mass of one kilogram an acceleration of one meter per second per second. In the MKS system the Newton is a unit of force and since weight is a force, the Newton is a unit of weight. As discussed in the previous paragraph, the weight of an object is related to the force of gravity acting on it. Thus, the weight of a 5.5 kg mass on the surface of the earth is equal to the product of the object's mass and the acceleration due to gravity at that place on the earth.

$$W = mg$$
$$W = (5.5kg)(9.8m/s^2)$$
$$W = 53.9kg \text{ X m/s}^2$$
$$W = 53.9N$$

TRY THIS EXPERIMENT:

fig. 11

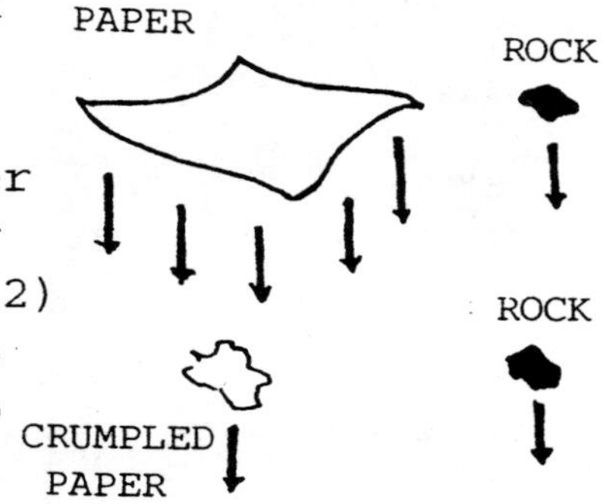

Try holding a sheet of paper and a small rock about four feet from the floor and allow both to drop to the floor at the same time. Notice that the rock falls first and the paper floats down (fig. 11). Now crumple the paper into a ball and repeat the experiment (fig. 12) This time both the paper and the rock hit the floor at the same time. According to Galileo the paper, crumpled or not, and the rock will fall at the same rate in the absence of air resistance. In his famous experiment at the Leaning Tower of Pisa, he simultaneously released two spheres, a heavy sphere made of iron and a lighter sphere made of wood. Inspite of the differences in weight, both spheres, dropped side by side, hit the ground at almost the same time. Galileo concluded that in the absence of air resistance all falling objects accelerate at the same rate.

PAPER ROCK

CRUMPLED PAPER ROCK

fig. 12

TERMINAL VELOCITY

However, there is air resistance and it affects the velocity of a falling object. If the object falls a great distance, the upward force caused by air resistance can equal the downward force caused by gravity. The net force is then zero and the object no longer increases its velocity as it continues to fall. It is said to have reached its maximum velocity which is called "terminal velocity". In other words, the object continues to fall but it is no longer accelerating. In a vacuum, the net force is only due to the weight force. If there is no air then there is no upward force and consequently all objects accelerate downward at the same rate. In air, the net force is the difference between the weight force and the air resistance. The force of air resistance acting on a falling object depends on the shape and size of the object (affecting the amount of air the object encounters) and the speed of the falling object. The acceleration is the same for heavier and lighter objects in free fall where there is no air resistance. A heavier object in free fall has a greater weight force acting on it, but it also has a greater mass which means that the ratio remains constant. This constant is equal to the acceleration due to gravity.

The force of air resistance reduces the downward net force on the object, thus reducing the acceleration. When the terminal velocity is reached, the acceleration is zero and the object continues to fall with a constant velocity. Zero acceleration does not mean zero velocity. It means the object will maintain its velocity and continue to fall without speeding up or slowing down.

NEWTON'S THIRD LAW OF MOTION

LAW III (THE LAW OF ACTION AND REACTION):

WHENEVER ONE OBJECT EXERTS A FORCE ON A SECOND OBJECT, THE SECOND OBJECT EXERTS AN EQUAL AND OPPOSITE FORCE ON THE FIRST OBJECT.

For every action, there is an equal and opposite reaction. Actions always occur in pairs. When you push an object and no motion takes place, then you feel a resistance which is the force the object exerts on you. Remember, according to Newton's Second Law, if the net force is not equal to zero, the object will move in the direction of the net force. When the net force is zero, then one force, the action force, is equal to the other force, the reaction force. The action force and the reaction force act on different objects. If object A exerts a force on object B, then object B exerts an equal and opposite force on object A.

Again, the forces occur in pairs acting on different objects and every action force has a reaction force. When you push against a wall, the wall in turn pushes against you. Push against a wall with a force of 10N (fig. 13). Since the wall does not accelerate, the net force acting on it must be zero. This means the wall is pushing you with a force of 10N. Suppose there is a book resting on a table (fig. 14). Since the book is not moving, the net force is zero and there is no acceleration. The table is preventing the book from reaching the floor. This means the table is exerting a force on the book equal and opposite to the force or weight force of the book.

fig. 13

fig. 14

Sometimes it is not easy to identify the reacton force. For example, the Earth exerts a gravitational force on each individual. At the same time the individual exerts an equal and opposite force on the Earth. Since the Earth is much more massive, its acceleration is so small and goes unnoticed. This idea is easier to understand with a rifle and bullet. When the bullet is fired, there is a reaction force on the rifle. The action force and the reaction force are equal. This means that the acceleration of the bullet is greater than the acceleration of the rifle because the mass of the bullet is less than the mass of the rifle. There is a kickback of the rifle but the acceleration of the bullet is so much greater (fig. 15).

$$a_{(rifle)} = \frac{F}{M} \qquad A_{(bullet)} = \frac{F}{m}$$

fig. 15

Getting out of a rowboat which is not tied down to a dock is not easy because the boat is not pushing back on your feet when your feet push the boat (fig. 16).

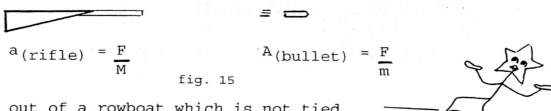
fig. 16

Once again, the action and reaction forces of Newton's Third Law are always acting on different objects, equal in magnitude and opposite in direction. Newton's Third Law is sometimes used to describe human behavior and is often referred to as getting even.

Sometimes, an "uncalled for" remark may bring a reaction of a negative response. By the same token, a compliment usually gets a very favorable reaction. A remark from someone usually elicits an equal remark from the other person. If you are nice, they are nice. If you are nasty, they are nasty. Also, you cannot have a good verbal fight with someone who will not answer back. But sometimes you may get more than an equal reaction because things on the other persons' mind may bring an overreaction. Most of the time, however, human behavior is one of responding in like manner which means that how you act will in some way determine how you are treated. A good deed is usually rewarded with another good deed. People usually respond with an equal reaction.

EXAMPLES FOR THOUGHT III

1. Which two quantities are measured in the same units?
   (1) velocity and acceleration (2) weight and force
   (3) mass and weight   (4) force and momentum

2. A table exerts a 2.0N force on a book lying on the table. The force exerted by the book on the table is
   (1) 20N   (2) 2.0N   (3) .20N   (4) 0 N

3. An object with a mass of 2 kilograms is accelerated at $5m/s^2$. The net force acting on the mass is
   (1) 5N   (2) 2N   (3) 10N   (4) 20N

4. What is the weight of a 5.0 kg object at the surface of the earth?
   (1) 5.0 kg   (2) 25N   (3) 49N   (4) 49 kg

5. The acceleration (in miles per hour per second) of an automobile that increases its speed from 20 miles per hour to 30 miles per hour in 1 second is
   (1) 30   (2) 35   (3) 10   (4) -10

6. The force required to give a 10 kg mass an acceleration of $5m/s^2$ is
   (1) 10N   (2) 5N   (3) 15N   (4) 50N

7. Suppose you are on a plane which is already in

motion and you are standing in an aisle. You decide to jump up as high as you could, but before you jump you draw an outline of your feet on the floor. When you come down and your feet land on the floor, will they be on the footprints you outlined, in front of the footprints or in back of the footprints? Why?

MATH HUMOR III

1. Prove: 2X10 = 2X11

2. Prove 101010 = 950 by adding one line and 2 dots

3. Where do you put "Z", above the line or below the line?

A   EF HI KLMN   T VWXY
_____
BCD G J   OPQRS U

Chapter IV

What would happen if the character at the
right let go of the string he is whirling
around with the ball on the end (fig. 1)?
Would the ball fly radially out, that is,
in line with the radius of the circle he is
creating or would the ball fly off in a path
tangential to the circle? A tangent is a
straight line which touches a circle at one
and only one point no matter how far it is
extended in either direction. What happens
when he lets the ball slow down? The answers
to these questions will become more apparent
in this chapter.

fig. 1

KEPLER'S LAWS OF PLANETARY MOTION

Kepler studied planets millions of kilometers away. Using
information collected by an astronomer named Tyco Brahe, Kepler
developed three laws of planetary motion.

LAW I: EACH PLANET MOVES IN AN ELLIPTICAL ORBIT WITH THE SUN AT ONE FOCUS.

Kepler found that the paths of the
planets are not circular but oval in
shape and that each planet goes around
the Sun in a curve called an ellipse
(fig. 2). The curve is such that at
any point on the curve the sum of the
distances from two fixed points is
constant (fig. 3). These fixed points
are called foci (plural for focus). With respect to the planets,
the Sun is at one focus and nothing is at the other focus.

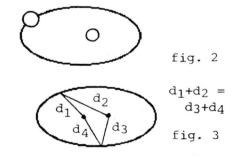

fig. 2

$d_1+d_2 = d_3+d_4$

fig. 3

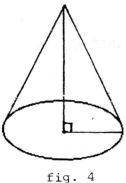

fig. 4

The circle, ellipse, parabola and the hyperbola
are sometimes referred to as conic sections
because   they could be described by using
the cone shown in the diagram at the left
(fig. 4). Note that the cone has a circular
base and that the line going through the center
of the cone is perpendicular to the radius of
the circle at the base.

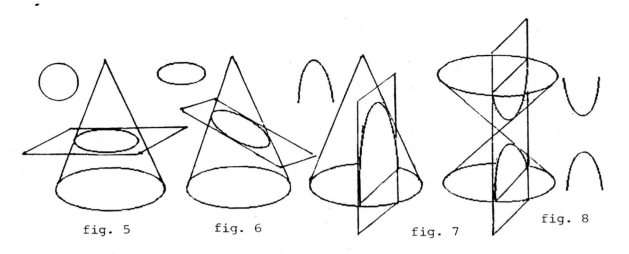

fig. 5 fig. 6 fig. 7 fig. 8

The circle, (fig. 5), the ellipse, (fig. 6), the parabola, (fig. 7), and the hyperbola (fig. 8) can be illustrated by taking different cross-sections of the right circular cone (fig. 4). These curves are part of our every day lives. The path of a baseball thrown between two players very closely resembles the parabolic curve (fig. 7). The cross-section of an automobile headlight (parrallel to the light beam) is also an illustration of the parabolic curve. The path of the electrons in the model of an atom is an illustration of an elliptical curve (fig. 6). The ellipse is also used in certain whispering galleries and is used in the famous Taj Mahal where a vertical cross-section of the dome reveals half of an elliptical curve. Suppose a person is standing at one focus and another person is standing at the other focus, several feet away. A remark whispered by one of these people would be heard by the other person, but the whispered remark would be inaudible to anyone standing elsewhere. Finally, an example of an hyperbola that can be seen in your every day life is in the reflection on the wall of a lighted lamp with a right cylinder lamp shade. There are two shadows on the wall, one at the top and one at the bottom which together form the curve known as an hyperbola (fig. 9).

fig. 9

LAW II: THE LINE FROM THE SUN TO ANY PLANET SWEEPS OUT EQUAL AREAS OF SPACE IN EQUAL TIME INTERVALS.

Kepler found that there is a relationship between a planet's speed and its distance from the Sun and as the planet moves around the Sun it does not move at a uniform speed. The planet moves faster when it is nearer to the Sun and slower when it

is farther from the Sun. This
change in speed is what Kepler refers
to in his second law which states that
if a line is drawn from the Sun to a
planet, it would sweep out equal
areas in equal time intervals (fig. 10).
This means that in order for the
triangular shaped regions to be equal
in area the planet would have to move
faster at times and slower at other
times for equal time intervals. The
time interval in going from A to B is
the same as the time interval in going
from C to D. However, the distance from
A to B is greater than the distance from C to D. If a planet
moves through a greater distance in the same amount of time, then
it must move faster. Notice that when the planet is closer to
the Sun (fig. 11), region I which it sweeps out is wider and thus
the planet's traveling distance is greater. This means that the
planet must move faster to cover the greater distance in the
given time interval. Conversely, when the planet is further from
the Sun (fig. 11), region II which it sweeps out is narrower and
the planet's traveling distance is less. In this case, the
planet can travel more slowly to cover this shorter distance in
the same time period. The implication is that the planets do not
move at uniform speeds.

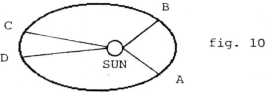

fig. 10

the time to go from A to B equals
the time to go from C to D
the distance from A to B is greater
than the distance from C to D

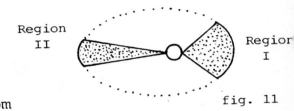

Region II

Region I

fig. 11

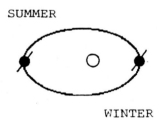

SUMMER

WINTER

fig. 12

It is interesting to note that the Earth is
farther from the Sun in the summer and closer
to the Sun in the winter (fig. 12). However,
because of the tilt of the Earth's axis, the
northern hemisphere is warmer in the summer
when the Earth is farther away from the Sun
and colder in the winter when the Earth is closer
to the Sun.

**LAW III: FOR ALL PLANETS THE RATIO OF THE CUBES OF THEIR AVERAGE
DISTANCES FROM THE SUN TO THE SQUARES OF THE PERIOD
OF REVOLUTION IS CONSTANT.**

In other words, for any planet in the solar system, the cube of
its average distance from the Sun divided by the square of the
period of revolution (the time taken by a planet to make one

complete trip around the Sun) is the same. When two quantities
are divided, this relationship can also be called a ratio. This
ratio is the same for all planets which is to say the ratio is
constant.

$$R^3/T^2 = K$$
$$R = \text{average distance from the Sun}$$
$$T = \text{period of revolution}$$
$$K = \text{constant}$$

This law applies to any system of satellites. Whenever a group
of smaller bodies orbit about a much more massive object, such
as the moons of Jupiter, R^3/T^2 is also constant for this system
but the value of the constant is different and it depends upon
the mass of the central body. This third law could also be
explained by examining the following table.

PLANET	R	T	R^3/T^2
	MEAN RADIUS OF ORBIT (METERS)	PERIOD OF REVOLUTION (SECONDS)	m^3/s^2
	AVERAGE DISTANCE FROM THE SUN	TIME NEEDED FOR ONE TRIP AROUND THE SUN	RATIO
Mercury	5.79×10^{10}	7.60×10^{6}	3.35×10^{18}
Venus	1.08×10^{11}	1.94×10^{7}	3.35×10^{18}
Earth	1.49×10^{11}	3.16×10^{7}	3.35×10^{18}
Mars	2.28×10^{11}	5.94×10^{7}	3.35×10^{18}
Jupiter	7.78×10^{11}	3.74×10^{8}	3.35×10^{18}
Saturn	1.43×10^{12}	9.30×10^{8}	3.35×10^{18}
Uranus	2.87×10^{12}	2.66×10^{9}	3.35×10^{18}
Neptune	4.50×10^{12}	5.20×10^{9}	3.36×10^{18}
Pluto	5.9×10^{12}	7.82×10^{9}	3.34×10^{18}

Kepler did not have any knowledge of the planet Pluto. The
informaton for Pluto was filled in later.

Newton believed that some force must be working on the planets,
otherwise, their paths would be straight lines. From Kepler's
second law, Newton saw that this force was the Sun. It is the
gravitational attraction between the Sun and the planets that
causes the planets to move in curved paths. Newton believed that
without gravity each planet would continue to move in a straight
line. With gravity each planet falls away from this straight

line toward the Sun and thus follows a curved path around the
Sun. Compare this concept to the opening example in this
chapter. A ball was being swung around on a string. The string
can be thought of as gravity keeping the ball moving in a
circular path. If the string is cut, it is similar to removing
the pull of gravity and the ball would go off on a straight path
tangential to the circular path in which is was traveling.
Remember, a tangent is a straight line which touches a circle at
one and only one point no matter how far it is extended in either
direction. In other words, the
ball wants to go straight but the pull of the
string keeps it going in a circle (fig. 13).
Newton also showed that gravitational attraction
is a mutual force between all objects. If the
Moon is pulling on the Earth, the Earth is
pulling on the Moon. In addition, as the Earth
pulls on any object, then the object pulls on
the Earth as well.

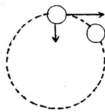

fig. 13

Kepler was able to predict the position of the planets, but the
explanation for their positions and paths was formulated by
Newton in his Law of Universal Gravitation.

NEWTON'S LAW OF UNIVERSAL GRAVITATION

Any two objects in the Universe attract each other with a force
which is directly proportional to the product of the masses of
the two objects and is inversely proportional to the square of
the distance between them. In other words, this statement says
that a mutual gravitational attraction exists between any two
objects and is expressed by the following formula:

$$F = \frac{Gm_1m_2}{d^2}$$

where m_1 = mass of the first object
            (read as m sub 1)
      m_2 = mass of the second object
            (read as m sub 2)
      d = the distance between the masses
      d^2 = the distance between the masses
            squared
      G = the constant of proportionality
            and is known as the universal

gravitation constant
$$G = 6.67 \times 10^{-11} \, N \cdot m^2/kg^2$$

(this value of G was determined by
another scientist many years later)

What does the exponent -11 mean? What does 10^{-11} equal?

10^1 equals 10
10^2 means 10X10 which equals 100 (a 1 with two zeroes)
10^3 means 10X10X10 "   " 1,000 (a 1 with three zeroes)
10^4 means 10X10X10X10 " 10,000 (a 1 with four zeroes)

This pattern is for positive exponents of ten.

Now, here is an explanation for the negative exponents of ten

10^{-1} means 1/10 which equals .1 (counting from the right of
                                     the one the dec. is 1 place
                                     to the left)

10^{-2} means 1/10X10 "   " .01 (dec. is 2 places to the left)
10^{-3} means 1/10X10X10 "   .001 (dec. is 3 places to the left)
10^{-4} means 1/ 10X10X10X10" .0001 (dec. is 4 places to the left)

Therefore,

10^{-11}= .00000000001 (dec. is 11 places to the left)

This is a very small number. It means that the gravitational
attraction between ordinary objects is usually too small to be
detected. Gravitational attraction becomes a significant force
when at least one of the objects is very large like the Earth.
The greater the masses, the greater the force of attraction
between the objects. The greater the distance between them, the
weaker the force of attraction. In the formula as the
denominator of the fraction gets larger then the fraction gets
smaller. For example, if the distance between the objects is
doubled, the force of attraction is reduced by 1/4. If the
distance is tripled, the force is reduced by 1/9.

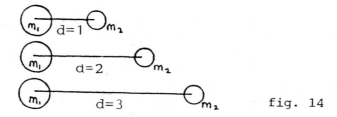

fig. 14

Assume $m_1 \times m_2 = 1$, that is, that the product of the masses is equal
to 1 (fig. 14).

Now, if the distance = 1, then the force of attraction

$$F = \frac{Gm_1 Xm_2}{d^2} = G\left(\frac{1}{1 \times 1}\right) = G(1)$$

If the distance = 2, then the force of attraction
$$F = \frac{Gm_1 X m_2}{d^2} = G\left(\frac{1}{2^2}\right) = G\left(\frac{1}{4}\right) = \frac{G}{4}$$

If the distance = 3, then the force of attraction
$$F = \frac{Gm_1 X m_2}{d^2} = G\left(\frac{1}{3^2}\right) = G\left(\frac{1}{9}\right) = \frac{G}{9}.$$

(If the numerator remains the same and the denominator increases the fraction decreases.)

In summary, there is a mutual force of attraction between any two objects, but as the objects move farther apart the force of attraction decreases. The force of attraction between a satellite and the Earth decreases as the satellite moves into space. Although this gravitational force of attraction never reduces to zero, the satellite may move so far that it is pulled into the gravitational field of another planet. In addition, the force of attraction exists between an individual and the Earth, between the Earth and the Sun, and even between two books.

On the other hand, as two objects move closer, the gravitational force of attraction increases. The closer the objects get, the greater the force of attraction. When an object falls towards the Earth the gravitaional force of attraction is increasing causing the object to fall faster. If the object is moving faster, then it is increasing its speed and, therefore, it is accelerating.

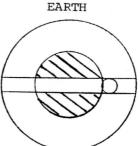

EARTH

fig. 15

Thus far, the gravitational field referred to is the field that is external to the Earth. This distance is measured from the center of the Earth. But the gravitational field extends to the interior of the Earth as well. For the gravitational field in the interior of the Earth, the mass of the Earth considered is only that portion of the Earth whose radius is the distance from the center of the Earth to the other object (fig.15). The other object being considered is theoretically in the Earth's interior. At the center of the Earth, the gravitational force on an object is zero. This means that the maximum gravitational force is at the surface of the Earth because the force of attraction decreases as an object moves into the Earth's interior and also decreases as an object moves away from the Earth.

It is interesting to note that there are other forces in the world. Forces such as electricity and magnetism have repelling as well as attracting forces, but gravity only attracts.

Since the Sun is much more massive than the Earth, one might wonder why the Sun doesn't pull an individual off the Earth and on to the Sun? According to Newton's Law of Universal Gravitation, the greater the distance between the objects the less force between them. Also, with smaller distances the force of attraction increases. Since the individual is much closer to the Earth than to the Sun, the force between the individual and the Earth is greater than the force between the individual and the Sun.

TIDES

The Earth pulls on the Moon causing the Moon to stay in orbit. The Moon pulls on the Earth and the Earth's bodies of water, causing the water to rise, giving the Earth tides. The Moon pulls more on the side of the Earth nearest it. The water nearest the Moon is pulled with more force than the Earth's center causing the water to produce a tidal bulge. In addition, the Earth's center is pulled with more force than the water on the farthest side causing a second tidal bulge. This produces two high tides and the depression between the bulges produces two low tides. The Sun also affects the tides but has less of an effect. When the Sun and the Moon are on the same line with that of the Earth, the high tides are higher than usual and the low tides are lower than usual. These higher high tides and lower low tides occur when there is a full Moon or a new Moon at which time there is the added pull of the Sun and they are called "spring" tides. This has nothing to do with the season of spring.

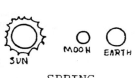

SPRING TIDE

NEAP TIDE

fig. 16

When the Sun's pull is perpendicular to the Moon's pull there are lower high tides and higher low tides and they are called "neap" tides (fig. 16). Because the Earth is not rigid, there are Earth

tides as well as water tides. If there are Earth tides and water
tides, what about the air? The atmosphere does experience tides
which is said to affect the behavior of living things. It has
been alleged that people behave differently during a full Moon
(spring tide) which might be related to the term "lunatic".

EINSTEIN

According to Einstein, gravitation is a distortion of space.
Imagine space as a lattice of rubber bands with two different
size balls on it making a depression in the lattice (fig. 17).
The further they are apart the less they affect one another. The
closer they are the more they will gravitate toward each other.
Einstein refers to this as the warping of space. Another analogy
to describe this concept is to envision a waterbed with two
individuals on it. The further apart they are, the less likely
they are to gravitate or to be pulled toward each other. And
conversely, the closer they are the more likely that they will
gravitate toward each other. Classical physics, as explained by
Newton's approach, holds true if the velocities of the objects
being studied are small compared to the velocity of light where
Einstein's theory is more applicable. Einstein's theories agree
with Newton's as mentioned when the speed is small compared to
that of the speed of light or when there is a weak gravitational
field.

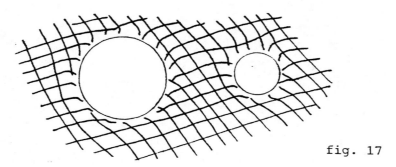

fig. 17

WEIGHTLESSNESS

The feeling of weightlessness is experienced by astronauts in an
orbiting spacecraft because the force of gravity keeps the
spacecraft falling toward the Earth. The spacecraft is falling
around the Earth and everything inside the spacecraft is falling
at exactly the same speed as the spacecraft. The floor is
falling at the same speed with which the astronaut is falling

which gives the astronaut the feeling of weightlessness. To keep
from floating the astronaut straps himself down or makes use of
the velcro on his space suit or space boots and on the velcro
mounted in the space cabin. There is no such thing as drinking
from a glass or cup. On earth gravity keeps water in a glass and
makes it pour from from the glass when the glass is turned upside
down. However, in a spacecraft, water stays inside a glass of
water which is turned upside down because everything is falling
toward the earth at the same rate. The astronaut must use a straw or
plastic squeeze bottle and he must eat bite sized food because
crumbs will float around the cabin.

To better understand the feeling of
weightlessness, imagine being in an
elevator where the chain snapped
(fig. 18). Further imagine you had
an apple in your hand and you let go
of the apple when the elevator began
falling. Since you, the elevator and
the apple would be in freefall, it would
appear as though the apple is floating
in air. It looks like the apple is not
falling because you, the elevator and
the apple are falling at the same rate.
If you were an observer on the outside

fig. 18

viewing someone else in the elevator, you would see the
individual, the elevator and the apple all falling at the same
rate. The apple would not be suspended in air.

TRY THIS EXPERIMENT:

fig. 19

Get a paper cup and make a small hole in
the bottom. Fill the cup with water and
watch the water drip from the bottom (fig.19).
Raise the cup high over a sink and then let
it go. As soon as the cup starts to fall
notice that the water stops dripping. This
is because the cup and the water are falling
at the same rate (fig. 20).

GRAVITATIONAL ACCELERATION

The Moon is a natural satellite constantly falling to the Earth
while it is in orbit. Instead of moving tangentially, every

second it falls toward the Earth away from the tangent line.
Just like a satellite, the Moon experiences what is called
gravitational acceleration. Newtown believed that the
gravitational acceleration of the Moon should be less than the
9.8m/s^2 experienced at the surface of the Earth because
acceleration falls off with increasing distance and the Moon is
quite far from the Earth. He felt that the Moon is falling
toward the Earth with a gravitational acceleration that is
significantly less than the 9.8m/s^2. Newtons's Law of Universial
Gravitation verified what he believed.

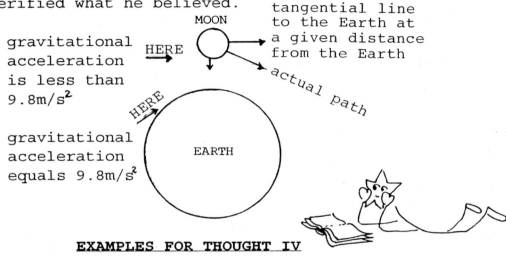

EXAMPLES FOR THOUGHT IV

1. When a plane that is parallel to the axis of a cone slices
    the cone, the cross section is a parabola. Describe how
    the plane must be positioned to produce a cross section
    that is a straight line?

2. Higher high tides and lower low tides caused by the added
    pull of the Sun at the times of a new and full Moon are
    called
    (1) neap tides   (2) summer tides   (3) spring tides (4) Sun tides

3. Two small objects of equal mass are at a fixed distance
    apart. How would the gravitational force between them change
    if the mass of each object were three times as great?
    (1) 3 times   (2) 9 times   (3) 1/3 as much (4) no change

4. As two objects move farther apart the gravitational force
    between them
    (1) increases   (3) remains the same
    (2) decreases   (4) not enough information

5. What is the force of attraction between two objects 4.0X10m apart if their masses are 4,800kg and 21,000kg, respectively?

6. What is the gravitational attraction between a 50kg girl and a 40kg girl who are 4m apart?

MATH HUMOR IV

1. Pick a number between 1 and 10.
   Multiply the number by 9.
   Subt. 5.
   If your answer has two digits, add them.
   If your answer still has two digits, add them again until
      you have one digit.
   Set 1 equal to A, 2 equal to B, etc.
   Think of a country in Europe that begins with that letter.
   Think of the second letter of the country's name.
   Think of a large animal whose name begins with that letter.
   Think of the color of that animal.
   ARE YOU THINKING OF A GREY ELEPHANT FROM DENMARK?

2. $(x-a)(x-b)(x-c)...(x-z) = ?$ (Find the product.)

3. How many seconds are there in one year?

4. $$\sqrt{-1} = \sqrt{-1}$$
   $$\sqrt{\frac{1}{-1}} = \sqrt{\frac{-1}{1}}$$
   $$\frac{\sqrt{1}}{\sqrt{-1}} = \frac{\sqrt{-1}}{\sqrt{1}}$$
   $$\sqrt{1}\sqrt{1} = \sqrt{-1}\sqrt{-1}$$
   $$1 = -1$$

   Find the fallacy.

Chapter V

PROJECTILE MOTION

A projectile is any object that is put into motion by some force
and continues in motion by its own inertia. If one projectile is
shot horizontally and an object is dropped vertically, both will
hit the ground at the same time. Gravity acts on both objects at
the same time and at the same rate. The horizontal velocity is
independent of the vertical pull of gravity.

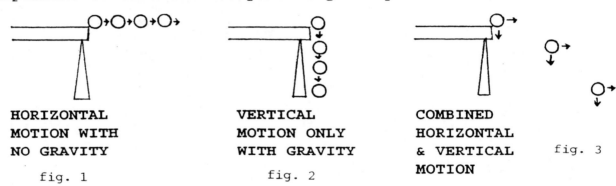

HORIZONTAL
MOTION WITH
NO GRAVITY

fig. 1

VERTICAL
MOTION ONLY
WITH GRAVITY

fig. 2

COMBINED
HORIZONTAL
& VERTICAL
MOTION

fig. 3

Thus, a package dropped from a moving airplane has two velocities
independent of each other. There is the continuously increasing
velocity caused by the pull of gravity and there is the constant
velocity imparted to it by the airplane. The actual path of
flight of the package is a half of a parabola. The forward
motion of the package can be illustrated by comparing the shadows
of the airplane and the package (see chapter III). It is because
the forward motion continues independent of the vertical motion
that planning must be done to insure that a package dropped from
an airplane will land in the appropriate area. If planning is
not done and the package is dropped above the desired area it
will fall beyond that area.

EARTH SATELLITES

From the above discussion it follows that if a ball is thrown
horizontally, the horizontal velocity remains constant
(neglecting air resistance) while only the vertical component of

the ball's velocity is changing. The force of gravity is causing
the vertical velocity to keep changing and consequently the ball
moves in a curved path. The greater the horizontal velocity the
greater the radius of the curved path (fig. 4). If the
horizontal motion is fast enough and the radius of the ball's
curved path is equal to the radius of the Earth, then the ball
would follow the curvature of the Earth and with no air
resistance it would continue forever. The following paragraph
uses the above information to discuss what is required for a
satellite to be placed into orbit around the Earth.

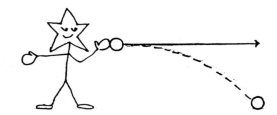

fig. 4

If a satellite is elevated and placed
into horizontal motion at 8km/s, it
falls 4.9m every second. Gravity is
pulling at right angles to the path
of the satellite. Earlier in this
discussion it was pointed out that
if an object moves horizontally,
gravity did not affect the speed of

the object. Gravity
is pulling perpendicular (at right angles) to
the direction in which the satellite is moving.
A tangent to a circle is a line which touches
a circle in one and only one point no matter
how far it is extended in either direction.

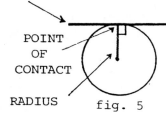

It is perpendicular to the radius
of the circle at the point of contact called the point of
tangency (fig. 5). The horizontal direction of the satellite is
tangent to the circle whose radius is the given
distance from the center of the Earth. When the satellite is
raised to the given altitude and put into motion horizontally
with a speed of 8km/s, it follows the curvature of the Earth's
surface at the given distance from the Earth (fig. 6). Each
second the satellite is falling toward the Earth because of the

force of gravity. The gravitational force is perpendicular to the horizontal motion which means that the horizontal speed remains at 8km/s. Without the force

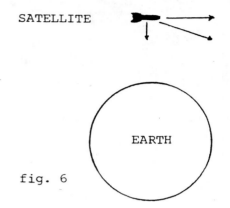

SATELLITE

EARTH

fig. 6

of gravity the satellite's path would be tangential to a circle at the given distance from the Earth's surface. But at the same time, gravity does exist and is causing the satellite to fall away from this tangential line of direction. The tangential speed is constant which means the satellite will be moving just as fast as it was initially.

What happens if a satellite is placed into orbit with an initial speed that is greater than 8km/s? It would be going too fast so that at the end of the first second the arc that it createss would not follow the curvature of the Earth. Gravity is no longer pulling at right angles. The pull of gravity must be perpendicular to the motion of the satellite in order for it not to affect the satellite's speed. If the pull of gravity is not perpendicular then the satellite's speed will either increase or it will decrease. As the satellite curves away from the Earth the force of gravity continually slows it down to a point where the satellite's path curves back to the Earth gaining the speed it lost. This is a cycle that keeps repeating and the satellite is then moving in an elliptical path (fig. 7).

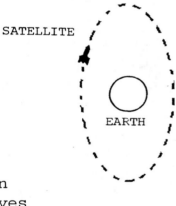

SATELLITE

EARTH

fig. 7

If the satellite was put into orbit with an initial speed that is less than 8km/s, then the satellite would fall back into the Earth as it attempts to circle the Earth (fig. 8). Remember holding a ball at the end of a string and trying to twirl it. If the proper speed is not achieved and it is too slow, the ball will fall in towards your hand.

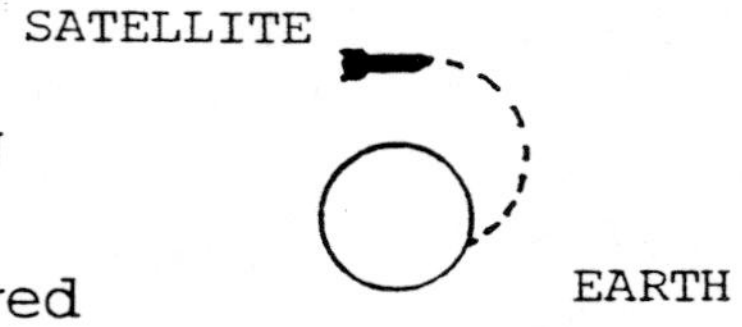

SATELLITE

EARTH

fig. 8

When the proper speed is achieved, the satellite is continually falling away from the tangent line. With the horizontal speed remaining constant and the fact that the satellite path follows the curvature of the Earth, the satellite should remain in orbit indefinitely. To put a satellite in orbit around the Earth, a rocket carrying the satellite is initially fired vertically so as to reach a certain altitude and then the last stage of the rocket is aimed horizontally propelling the satellite into a horizontal path with an initial speed of 8km/s. The satellite now continues to circle the Earth.

ESCAPE VELOCITY

A satellite put into orbit with an initial speed of 11km/s or more could never have its speed reduced to zero by the Earth's gravitation. According to the inverse square law $F = G(m_1 m_2)/d^2$, the force of attraction decreases with greater and greater altitudes and the satellite slows by smaller and smaller amounts every second. Although a satellite in orbit with a speed greater than 11km/s would never escape the Earth's gravitational field, it would be continually slowing down. It would not slow down enough to come back around. The satellite would continue to move away and may even be caught up in the gravitational field of another planet.

ROTATION VS REVOLUTION

The terms of rotation and revolution are frequently used when referring to orbiting objects and planets in space. If the axis about which the object is turning is inside the object, the object is said to be ROTATING (fig. 9). If the axis about which the object is turning is outside the object, the object is REVOLVING (fig. 10). For example, the Earth rotates about its axis daily and it revolves about the Sun yearly.

ROTATING   fig. 9   fig. 10   REVOLVING

THE SMILING FACE OF THE MOON:

We only see one side of the Moon from any point on the Earth. Cameras have taken pictures of the other side for us. We only see one side of the Moon because it rotates on its axis while it revolves around the Earth (fig. 11).

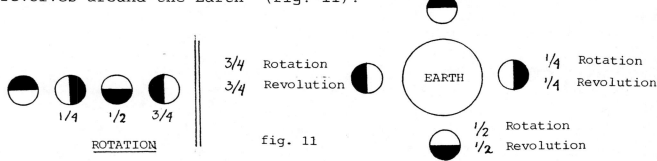

fig. 11

As the Moon makes one quarter of a revolution it is also making one quarter of a rotation. At this position the same side of the Moon is facing the Earth. This happens again at one half of a revolution and rotation (fig. 11). The Moon rotates on its axis taking 27 1/3 days to make a complete rotation. A complete revolution of the Moon around the Earth also takes 27 1/3 days. Since our calender month is longer, occasionally there will be two full Moons in one month instead of the usual occurance of one full Moon in a month. Years ago it was thought that the heavens seemed bluer than usual for the second full Moon and it was referred to as a blue Moon. Hence, the phrase "once in a blue Moon" became associated with a rare event. It is possible to have two extra full Moons in a one year period which would give us 14 full Moons in the year that this occurs. With the assistance of the computer it is relatively simple to calculate in advance each occurrance of a blue Moon.

ROTATIONAL INERTIA:

An object rotating about an axis tends to remain rotating about the same axis unless this is changed by some external force. The external force is called a TORQUE (pronounced like "fork"). Rotational inertia depends on the mass and also depends on the distribution of the mass with respect to the axis of rotation. The greater the distance between the bulk of the object's mass and its axis of rotation the greater the rotational inertia.

A young child walking on a railroad railing often stretchs his arms out to maintain his balance. If he distributes more of his mass further from the axis of rotation, he gains greater rotational inertia. A tight

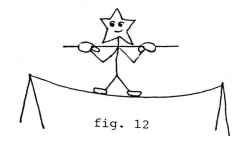

fig. 12

rope walker does the same thing when
he holds a bar across his sholders.
This reduces his tendency to rotate
(fig. 12).

TORQUE:

A torque can cause an object to rotate and can stop an object
from rotating. A push toward or away from the center of rotation
is not a torque. A torque is applied tangentially to the circle
of rotation. Torques produce changes in rotational motion by
either putting an object into rotational motion or stopping the
rotational motion.

A torque is defined as the product of the lever arm and the
applied force that tends to produce a rotation about an axis. It
produces a clockwise or a counter-clockwise rotation.

TORQUE = LEVER ARM X FORCE

LEVER ARM is the distance from the

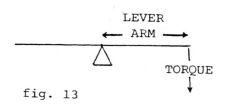

fig. 13

axis of rotation and is
perpendicular to the line
along which the force acts.
In other words, it is the
distance from the axis of
rotation (fig. 13).

On a seesaw a child learns that his distance from the axis of
rotation is as important as his weight. The torque caused by a
child on the right would tend to produce a clockwise rotation and
the torque caused by a child on the. left would tend to produce a
counter-clockwise rotation. The
diagram at the right (fig. 14), shows
that if the torques are equal there
is no net torque and no rotation is
produced. A greater torque can be
generated on one side by increasing
the length of the lever arm (the
distance from the fulcrum), by
increasing the applied force (the
weight force of the child) or by
doing both.

fig. 14

A torque is applied when turning the steering wheel of a car (fig. 15). The rotation of the steering wheel is acomplished by pulling tangentially to the radius of the steering wheel. Nothing would be accomplished by pulling radially outward.

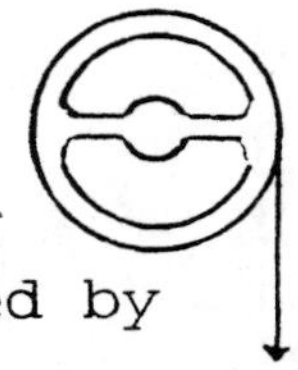

fig. 15

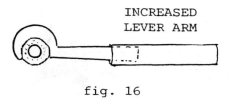

INCREASED
LEVER ARM

fig. 16

By slipping a piece of pipe over the handle of a wrench the lever arm is increased and a greater torque would result from the same force because the lever arm is longer (fig. 16).

The door knobs on doors are placed as far from the hinges to produce a greater torque and thus make the door easier to open. The hinges act as the fulcrum, the pivotal point about which a lever can rotate. Imagine how difficult it would be to open the door, if the door knob were on the side of the door near the hinges. Even the door knob itself is wide so that it is easier to turn (fig. 17).

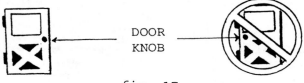

DOOR
KNOB

fig. 17

Archimedes, a Greek philoshper who lived in the first century B.C., was impressed with the power of the lever that he is said to have exclaimed, "Give me a place to stand and a fulcrum on which to rest a lever and I will move the Earth."

If you want to make an object move, apply a force. If you want to make an object spin, apply a torque. A torque produces rotational motion. The concept of rotational motion gives rise to the concept of the center of gravity and to the concept of stability. If an object is supported from its center of gravity and every object has a center of gravity, it will not rotate. That is, gravity will not produce a torque and the object will not rotate. To prevent rotational motion from occurring, the clockwise torques must equal the counterclockwise torques.

If you balance a pencil in a horizontal position using your finger tip as the fulcrum, and if there is no rotation, then the torque on each side is equal.

This point of a pencil is the center of gravity which may or may not be the geometric center (fig. 18). The weight of the pencil can be thought of as being at this point. Actually, the weight is the resultant of all parallel forces acting on the pencil and the resultant goes through the center of gravity.

fig.18

CENTER OF MASS AND CENTER OF GRAVITY

For a given object the center of mass is the average position of all the particles of mass that constitutes the object. A ball has a center of mass at the center. A cone has a center of mass one-fourth of the way up from its base. The center of mass of an object can be outside the mass of the object. In the case of a ring or a hollow sphere, the center of mass is the geometric center where no matter exists. The center of mass of a boomerang is also outside the mass which constitutes the boomerang.

TRY THIS EXPERIMENT:

Try hooking two forks together at the prongs with a coin caught between. Now try balancing these forks with the coin resting on the rim of a glass (fig. 19).

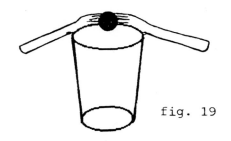

fig. 19

The center of gravity is the average position of the weight distribution. Since weight and mass are proportional to each other, the center of gravity and the center of mass refer to the same point of the object. However, for very large bodies such as the Moon, the center of gravity is not at the center of mass. Closely associated with torque and center of gravity is the stability of a rigid body. A rigid body is a system of particles in which the particles are fixed or constant distances apart. Water would not satisfy this criterion. But if water were frozen, the ice would be more like a rigid body.

STABILITY

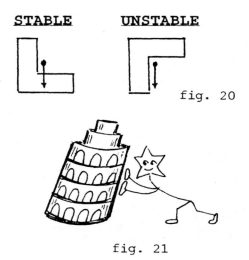

fig. 20

If a line were dropped straight
down from the center of gravity
of an object of any shape and if
it falls inside the base of the
object, the object is said to
be in stable equilibrium. If the
line falls outside the base, the
object is said to be unstable.
(fig. 20). A line from the center of
gravity of the Leaning Tower of Piza
falls within the base and, therefore,
the tower is stable (fig. 21).

fig. 21

In order to reduce the likelihood of tipping, an object should be
designed with a wide base and a low center of gravity. When
a person stands, his center of
gravity is somewhere above the
area outlined by his feet. If
his feet were spread farther apart,
making the base wider, this would
increase the area outlined by his
feet bringing greater stability
(fig. 22). When a person stands
against a wall and leans forward
his center of gravity extends
beyond the area outlined by his
feet. He loses his balance and
begins to fall (fig. 23). When
a person carries a bag of groceries
he leans backward keeping his center
of gravity above his feet because
the center of gravity is shifted
with the added groceries.

fig. 22

fig. 23

Vases with wide bases have
more stability than vases
with narrow bases. Large,
tall trucks are designed
with low centers of gravity
to give them more stability
(fig. 24).

UNSTABLE

STABLE

fig. 24

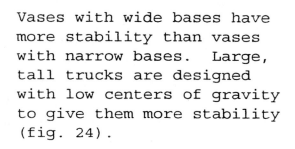

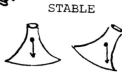

CENTRIPETAL FORCE VS CENTRIFUGAL FORCE

Centripetal force is any force that causes an object to follow a circular path. It is a center-directed force. Centripetal means "center-seeking" or "toward the center". It is any force whether a string tension, gravitational pull or electrical, etc., that is directed at right angles to the path of the moving object and produces circular motion. It plays the main role in the operation of the centrifuge. The centrifuge is a device for separating particles of varying density.

Centripetal force also plays a role in automatic washing machines. In the spin cycle the high speed produces a centripetal force on the clothes. The holes in the drum allow the water to escape and thus the centripetal force is not exerted on the water. The clothes are forced into a circular path away from the water and the water's path is a straight line tangential to the drum (fig. 25).

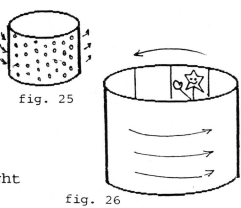

fig. 25

fig. 26

In some amusement parks there exists a ride where the floor drops out as the ride spins around (fig. 26). It is a common misconception to think that centrifugal force pins the riders to the wall as the ride spins around. Centrifugal force is considered to be an outward force and is defined as "center-fleeing" or "away from the center". There is no centrifugal force acting on the rider. The centrifugal force effect is not attributed to any real force but to inertia which is the tendency of a moving object to follow a straight line path. If the wall of the ride were not there the rider would move in a straight line tangent to the circular wall of the ride. The ride's wall is in the way and is the centripetal force which makes the rider follow a circular path. This is the same thing that happens when we hit against the side of a car as it turns a corner.

A satellite in orbit around the Earth is in free fall with centripetal acceleration due to gravity. The centripetal force or center-seeking force is provided by the gravitational force of attraction between the satellite and the Earth and is directed toward the Earth. Centripetal acceleration refers to an object moving in a circle where the velocity keeps changing. The word acceleration is required because the direction is constantly changing even though the speed may or may not change.

EXAMPLES FOR THOUGHT V

1. If we could fire an atomic cannon which shoots projectiles with the speed of light, how long would it take to hit the Moon?, the Sun?.

2. Why is it easier to balance yourself with your arms outstretched when walking on the rail of a railroad track or on a narrow wall or fence?

3. Describe the motion of a bullet fired from a gun at a height of 78.4m above the ground if the bullet leaves the gun at 100m/s.

4. There are two weights one at 275 pounds and the other at 125 pounds. They are at both ends of a 4" bar. Where should the fulcrum be placed so that they would be in equilibrum.

5. Look up the definition for centrifugal force in a scientific dictionary.

6. A satellite is moving a constant speed in a circular orbit about the Earth, as shown in the diagram below.

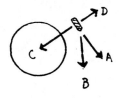

    The net force acting on the satellite is directed toward point?
    (1) A   (2) B   (3) C   (4) D

7. The centripetal force acting on the Moon is
    (1) a string   (3) the gravity of Earth
    (2) the Sun   (4) not enoungh information

8. If the gravitational force acting on a satellite were removed, then the satellite would move
   (1) radially outward   ✓(3) off on a tangent
   (2) radially inward   (4) none of these

MATH HUMOR V

1. What is in the center of gravity?

2. Suppose there is an old two volume encyclopedia set sitting on a shelf in the library. The books no longer have covers and each volume is one inch wide. Sitting beside volume one at page one is an inch worm which can eat through one inch of paper in one hour. How long will it take this inch worm to go from page one of volume one to the last page of volume two?

3. Why was 10 afraid of 7?

4. If it takes 12 minutes to cut a board into four equal parts, how many minutes would it take to cut the same board into five equal parts, cutting at the same rate?

5. Pick a card from an imaginary deck of ordinary cards.

        Double the number on the card
        Add 1
        Multiply by 5
        If it is a club,   add 6
        If it is a heart,   add 7
        If it is a spade,   add 8
        If it is a diamond,   add 9

   Now tell me the number you got and I'll tell you the card you picked. For example, if your result is 22, then you picked an Ace of hearts. How did I do that?

Chapter VI

MOMENTUM

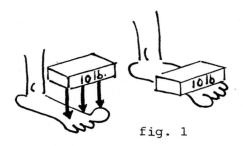

fig. 1

Suppose a 10 pound brick is placed on your foot. Now suppose a 10 pound brick is dropped on your foot (fig. 1). Which would hurt more? Why? Let's try another experiment. Suppose a 10 pound brick and a 50 pound brick are dropped on your foot. Now which will hurt more? Why? In answering the first question, notice that the difference between the two bricks is their velocities. The falling brick, having the greater velocity, would cause more pain. In the second experiment, notice that the difference is their masses and the falling 50 pound brick with the greater mass would do more damage to your foot. The combination of the object's velocity and mass is known as the object's momentum. The momentum of the dropped or falling 10 pound brick is greater than the stationary 10 pound brick and the momentum of the falling 50 pound brick is greater than that of the falling 10 pound brick. The answer to the question of "why" will be explained in this chapter.

MOMENTUM = MASS X VELOCITY

In a product, if either of the factors or both of the factors is increased, then the product increases. Also, if either of the factors or both of the factors is decreased, then the product decreases. Therefore,

1. the greater the velocity the greater the momentum.
2. the greater the mass the greater the momentum.
3. the greater the velocity and the mass the greater the momentum.

A truck has more momentum than a car going at the same speed because its mass is greater. Two cars of the same mass traveling at different velocities have different momenta. A truck at rest has no momentum because the velocity is zero. Even though momentum is a function of the velocity of an object, the momentum cannot be used to determine the object's velocity. Compare the momentum of a boat of mass 10^6 kg moving at 0.1m/s with that of a truck whose mass is 2×10^3 kg and moving at 50m/s

$$\text{Momentum of the boat} = mv$$
$$= (10^6 \text{kg})(0.1\text{m/s})$$
$$= 10^5 \text{kg} \times \text{m/s}$$
$$\text{Momentum of the truck} = mv$$
$$= (2 \times 10^3 \text{kg})(50\text{m/s})$$
$$= 10^5 \text{kg} \times \text{m/s}$$

The momenta are the same. If only the momenta were given, then it would not be possible to calculate either velocity without more information.

A train is much more difficult to stop than a car moving at the same velocity. The train has more momentum because its mass is greater than that of the car. A bullet of small mass may be given a large momentum if its velocity is made very great.

One method for understanding motion was presented by Newton in his three laws of motion. Momentum, on the other hand, describes motion in an alternative way. The method of using momentum to describe motion may be more convenient in certain situations. There are interesting relationships between the two methods and the following discussion presents some of these comparisons.

COMPARISON #1:

NEWTON'S LAW I: Law I deals with inertia.

MOMENTUM: Momentum is referrred to as inertia in motion.

COMPARISON #2:

NEWTON'S LAW II: Law II states $F = ma$.

MOMENTUM:

$$\text{Using } F = ma, \tag{1}$$
$$\text{and} \quad a = \frac{\text{change in velocity}}{\text{time}}$$
$$\text{then} \quad a = \frac{\text{final vel. - initial vel.}}{\text{time}}$$

Substituting for "a" in (1) we get
$$F = m \frac{(\text{final vel. - initial vel.})}{\text{time}}$$

Using the distributive law, $F = \dfrac{m(\text{final vel.}) - m(\text{initial vel.})}{\text{time}}$

Multiplying both sides of the equation by time (t = time) gives us

$$Ft = m(\text{final vel.}) - m(\text{initial vel.})$$
$$Ft = mv_{final} - mv_{initial} \qquad (2)$$

The right side of the equation is a change in momentum (the final momentum minus the initial momentum). This change in momentum is called IMPULSE. Since the change in momentum equals Ft (see line 2), and since impulse is the name for the change in momentum then it follows that:

$$\textbf{IMPULSE = Force X time}$$
$$= \textbf{Ft}$$

Impulse is the product of force times the time during which the force acts. An impulse applied to an object causes that object to change its momentum. If a moving car having momentum (its mass times its velocity) hits a wall and stops, its momentum changes to zero. The mass remains the same but the velocity equals zero (fig. 2). (When one factor of a product, such as "mv", is zero, the product is zero). If the same car hits a hay stack instead of a wall, it would also stop. But, it would take longer as the car moved into the hay stack. Since more time is involved less force is necessary to accomplish the same thing, that is, stopping the car. With less force there is less damage.

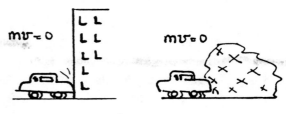

fig. 2

$$F \quad X \quad t = f \quad X \quad T$$

Greater x Smaller = Smaller x Greater
force   amount   force   amount
        of time   of time

The aim is to increase the contact time and to reduce the impact force. In order to bring the car to a halt you need to bring the momentum to zero. You could slam your foot on the brake pedal for an emergency stop. This produces the necessary impulse using a large force exerted over a short time interval. If time permits, you could gently brake to a stop also producing the necessary impulse using a small force exerted over a long time interval. Of course, the second method is preferable, if time permits. In one case the car is stopped instantly and in the other case the car is stopped gradually. In most situations, a change in momentum involves a change in velocity because the mass is usually constant.

Acrobats use safety nets because a net extends the time of stopping the acrobat thus reducing the force necessary to change the momentum of the fall (fig. 3). If the acrobat hits the hard floor the acrobat would be stopped more abruptly requiring a greater force and would probably cause more damage. In either event the acrobat is brought to a stop. (More time, less force or less time, more force). The same theory explains why jumping on a wooden floor is easier on an individual than jumping on a concrete floor, since the wood floor has slightly more give than the concrete floor. Bending one's knees while landing after a jump also extends the time needed to stop and thus reduces the force on the individual (fig. 4). Boxing gloves serve the same purpose. The cushioned gloves prolong the collision time thus reducing the force (fig. 5). Baseball players pull their arms inward after catching a ball also to increase the time necessary to stop the ball and thus decrease the force which may cause them harm (fig. 6).

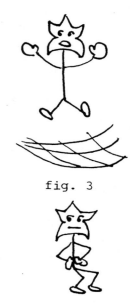

fig. 3

fig. 4

fig. 5

fig. 6

Understanding the concept of impulse is necessary to explain why the 10 pound brick dropped on your foot would hurt more than the 10 pound brick placed on your foot (fig. 1). The falling brick has momentum (mass times velocity) which is brought to zero when it hits your foot. A significant amount of force is exerted in the short time it takes to stop the brick. This explains why it could hurt. Compare this to the brick that is placed on your foot and already has zero momentum. The second part of the example states that two bricks are falling but one is 50 pounds and the other is 10 pounds. The 50 pound brick has a greater momentum because of its mass. It would also require a greater force to stop it because of the greater mass and therefore the greater momentum even though the stopping time is the same for both bricks.

<u>COMPARISON #3</u>:

<u>NEWTON'S LAW III</u>:   ACTION & REACTION

<u>MOMENTUM</u>:   In a closed system the total momentum is conserved.

1) A closed system means a system of colliding objects where the forces that act during the collision are not external forces such as a wall. An external force is required to change the momentum of an object. An internal force does not change the momentum. Pushing the dashboard in a car does not change the momentum of the car.

2) The term "conserved" means to remain unchanged or constant.

The force exerted on a bullet fired from a rifle is equal and opposite to the force exerted on the rifle (fig. 7). Before the gun is fired, the gun with the bullet has zero velocity and therefore zero momentum. Since no forces outside the gun are acting to accelerate either the gun or the bullet, it means that the total momentum after firing must remain zero. The forward momentum of the bullet must be equal and opposite to the momentum of the gun. The total net momentum remains zero.

RIFLE   BULLET

                                                                              fig. 7

−Mv ⟵   ⟶ mV

Velocity is a vector and, therefore, also shows direction. The rifle of greater mass has the smaller velocity and the bullet with a much lighter mass has a significantly greater velocity. The velocities of the bullet and recoiling rifle are not equal. The momentum is conserved not the velocities. If the masses of the bullet and rifle were the same then the velocities would also be the same but in opposite directions.

The idea of a rifle recoiling can be used to understand rocket propulsion. A rocket sends a satellite or spacecraft up into the atmosphere and the satellite circles the Earth. Imagine that each molecule of the exhaust gas is a bullet shot from the rocket. The total momentum of the rocket at any moment is equal to the total momentum of the ejected molecules of exhaust gas and is opposite in direction. Momentum given to the exhaust gases is countered with an

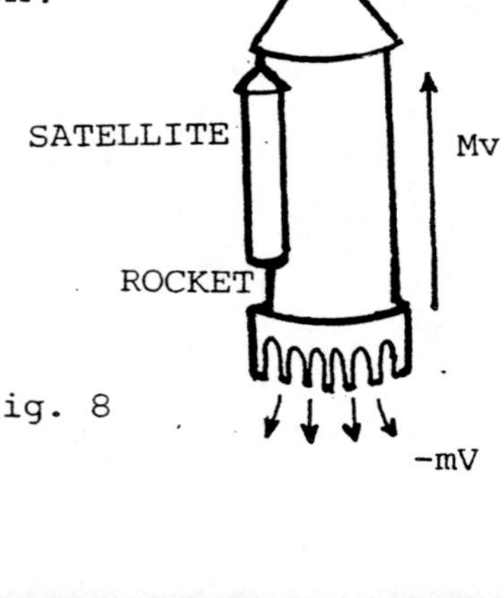

SATELLITE   Mv

ROCKET

fig. 8

−mV

equal and opposite momentum of the rocket
(fig. 8). The exhaust gas molecules are
ejected out of the rear of the rocket at
speeds of several thousand meters per
second and at a constant rate. Fuel is considered as part of the
mass of the rocket and as the fuel is ejected during lift off the
mass of the rocket decreases. A decrease in mass means an
increase in velocity. The increasing velocity is due to the
rocket's decreasing mass which could be further reduced by using
multistage rockets. The burned out stages are jettisoned off so
as to bring further reduction in mass. There are methods of
calculating how much of the mass of a rocket must be thrown away
to get an Earth satellite into orbit. By Newton's 3rd law, just
as a rocket exerts a force on its exhaust gases, the gases exert
a reaction force on the rocket.

A bazooka, or rocket launcher, used as a weapon against tanks,
consists of a tube which is open at both ends allowing the
exhaust gas to go out the back end. This produces very little
recoil. The exhaust blast which is necessary to project the
shell is quite powerful and can be dangerous if a person is
standing too close to the rear.

An astronaut approaching the Moon must fire retrorockets to slow
down the spacecraft so it will not crash into the Moon. The
force on the rocket causes a reaction in the opposite direction
on the spacecraft which slows down the spacecraft.
The diagram at the right shows five marbles
suspended from two rods by very fine nylon
strings (fig. 9). When one marble is pulled
out to the left and then released, one marble
swings out on the right side. When two
marbles are pulled to the left and
released, two marbles swing out on the
right. Repeating this procedure with
three marbles shows that three marbles
swing to the right. The middle marble

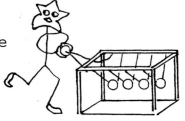

fig. 9

goes both ways. The reason for this is conservation of momentum.
The momentum entering on the left must equal the momentum exiting
on the right for one run from left to right. Read Chapter IX to
see why two marbles pulled and released from the left could not
result in one marble swinging out on the right side with twice
the velocity.

Collisions are classified as <u>inelastic</u> or <u>elastic</u>. An inelastic collision is characterized by deformation and/or generation of heat, sound, or light (sparks may fly). In general, there is a loss of some form of energy. In an elastic collision no deformation occurs and no heat is generated.

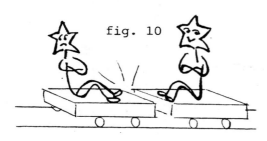
fig. 10

Suppose one railroad car is standing on the railroad track and another car of equal mass, moving from the left at 10m/s, locks with the stationary car and moves forward together with the other car (fig.10). Since the total momentum before the interaction is the same as the total momentum after the interaction, the two cars having twice the mass move at half the velocity of 5m/s. (Ignore friction and assume this is an elastic collision.)

Another example might be where the moving railroad car comes to a halt after the collision and the stationary car now moves to the right at 10m/s, keeping the total momentum constant before and after the interaction (fig.11).

fig. 11

It might also occur where both cars are moving toward each other, collide and then push off in opposite directions. The total momentum before the collision is equal to the total momentum after the collision (fig.12).

fig. 12

The rotary water sprinkler is another good example of the reaction principle describe by Newton's 3rd law or the law of conservation of momentum. The jet of water coming from the nozzle has a certain momentum. As a result the momentum on the nozzle is equal and opposite in direction. This causes the sprinkler to spin in the direction opposite to the stream of water.

The mathematics is more complex when dealing with cars that collide at various angles (not head on) and move apart at various angles to one another. Whatever the nature of the collision or however complicated the collision, the total momentum before, during and after remains the same.

ANGULAR MOMENTUM

Just as a mass moving in a straight line has linear momentum, a mass moving in a circular path has angular momentum. An example might be the planets orbiting the Sun or a rock whirling at the end of a string. The momentum depends on the mass "m", the speed "v", and the radial distance "r" from the axis about which the object is revolving and is equal to the product, "mvr" or "mass X velocity X radial distance".

An external net impulse is required to change linear momentum and an external net torque is required to change angular momentum. Newton's LAW OF INERTIA applies to linear momentum and also relates to angular momentum.

> "An object or system of objects will maintain its state of angular momentum unless acted upon by an unbalanced external torque."

There is no net torque acting on the Earth, therefore angular momentum is conserved. This means that for a given mass such as the Earth the product of "mvr" is constant. If "r" becomes smaller, then "v" becomes larger and if "r" becomes larger, then "v" becomes smaller. Notice that "m" is not changed because it refers to the mass of the Earth. In most instances a change in momentum involves a change in velocity or a change in radial distance since the object is the same and its mass remains constant. The Earth's velocity is greater when it is nearer to the Sun and slower when it is farther from the Sun (fig.13). This is in agreement with Kepler's second law about the planets sweeping out

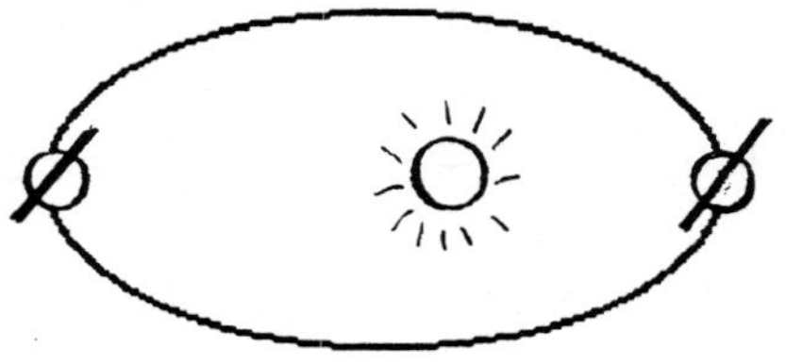

$$mvR = mVr$$

fig. 13

equal areas in equal time intervals. The planets speed up or slow down as they approach or recede from the Sun, keeping their angular momentum constant.

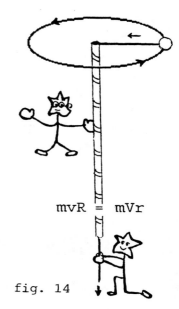

TRY THIS EXPERIMENT:

Try pulling a string through a straw with a ball tied to the end of the string. Now swing the ball at the end of the string around in a horizontal circle. Next pull the string down through the straw making the radius of the circle smaller (fig.14). Notice that the ball speeds up. No torque is being aplied because the pull of the string is a force along the radius and it takes a tangential force to produce a torque. The ball moves faster because the radius is smaller and the angular momentum is conserved.

$$mvR = mVr$$

fig. 14

The two previous examples dealt with the momentum of objects that revolve around a center axis. To repeat the earlier concept that for a particle of mass "m" moving in a circle, the angular momentum is defined as angular momentum = mvr, where "m" is the mass, "v" is the instantaneous tangential velocity and "r" is the radius of the circle. Calculations dealing with angular momentum are really not simple especially for rotating objects. For a rigid object, the formula for angular momentum changes. If an object is rotating on its axis, every atom of the object has its own mass, its own radial distance from the center and its own velocity. Rotational motion or angular momentum of a spinning object depends on (1) how fast the object is rotating, (2) the mass of the spinning object and (3) how the mass is distributed relative to the axis of rotation. Moment of inertia incorporates some of these principles. Moment of inertia is somtimes called rotational inertia.

Inertia resists changes in linear motion. Rotational inertia resists changes in rotational motion. Both types of inertia depend on the mass of the object, but rotational inertia also depends on the distribution of the mass about the axis of rotation. The total moment of inertia of a rotating object is found by adding up the moments of inertia of each small particle

of the object which includes each particle's mass and distance
from the axis of rotation. The rotational inertia or moment of
inertia for the object is equal to a constant times the object's
mass times the square of the radial distance from the axis of
rotation. The constant is a numerical value that

depends on the shape of the object and the location
of the axis about which the object is to be rotated.
Even the same object will have different moments
of inertia depending on the axis of rotation which
would cause the distribution of the mass to vary.
A stick rotated from one end will have a different
moment of inertia than a stick rotated from its
center (fig.15). In the first case, there are
more particles further from the axis of rotation
and in the formula for the moment of inertia, the
distance from the axis of rotation is squared.
Objects of the same size, same dimensions and the
same axis of rotation have different moments of
inertia if they are made of different material,
such as a steel door versus a wood door. This
is because their masses are different. Even two
wood doors of the same size and mass have different
moments of inertia if one is built with a window

fig. 15   in it.

Moment of inertia is the rotational comparison to mass which is
the measure of inertia. Objects with large moments of inertia
are more difficult to start or to stop rotating just as objects
with more mass are more difficult to start or to stop moving in a
straight line. Because of rotational inertia a solid cylinder
starting from rest will roll down an hill faster than a hollow
cylinder. The greater the distance between the bulk of an
object's mass and its axis of rotation the greater the moment of
inertia or the rotational inertia. The hollow cyclinder has its
mass farther from its axis of rotation than a solid cylinder. If
the moment of inertia is greater then there is more resistance to
change.

The concept of angular velocity and moment of inertia combine to
give the angular momentum of a rotating object. The angular
velocity for all particles in a rotating rigid object is the same
because they all go through the same angle, measured in radians,

in the same time interval. The spin angular momentum is equal to the moment of inertia times the angular velocity.

For rotating objects:
SPIN ANGULAR MOMENTUM = MOMENT OF INERITA X ANGULAR VELOCITY

Remember for revolving objects:
ANGULAR MOMENTUM = MASS X VELOCITY X (RADIUS OF PATH)

When an ice skater spins around on the ice and then draws her arms in, she spins even faster. This has to do with conservation of angular momentum. If there are no external forces, then the total angular momentum before she pulls her arms in is the same as the total angular momentum after she pulls her arms in. When she is spinning and her arms are extended, her mass is distributed farther from the axis of rotation giving her a greater moment of inertia than when her arms are pulled in. Remember, the angular momentuum is equal to the product of the moment of inertia and angular velocity. If the angular momentum remains constant and if the moment of inertia decreases (pulling in her arms) then the angular velocity increases. As a result, the skater spins faster. When the skater wants to stop she extends her arms, increasing her moment of inertia and thus decreases her angular velocity.

ANGULAR MOMENTUM   =   ANGULAR MOMENTUM

SMALL   LARGE   LARGE   SMALL
ANGULAR   X MOMENT   ANGULAR   X MOMENT
VELOCITY   OF   VELOCITY   OF
             INERTIA   INERTIA

fig. 16

EXAMPLES FOR THOUGHT VI

1. A 2.5kg toy cannon is at rest on a frictionless surface.
    A remote triggering device causes a 0.025kg projectile to

be fired from the cannon. Which equation best describes this system after the cannon is fired?
(1) mass of cannon + mass of projectile = 0
(2) speed of cannon + speed of projectile = 0
✓ (3) momentum of cannon + momentum of projectile = 0
(4) velocity of cannon + velocity of projectile = 0

2. A 3kg object traveling 10m/s north has a perfect elastic collision with a 6kg object traveling 5m/s south. What is the total momentum after the collision?
✓ (1) 0 kg X m/s   (2) 30kg X m/s north
(3) 30kg X m/s   (4) 60kg X m/s east

3. A 6.0kg mass is moving at 4.0m/s toward the right and a 8.0kg mass is moving at 3.0m/s toward the left on a horizontal frictionless table. If the two masses collide and remain together after the collision, their final momentum is
(1) 1.0kg X m/s   (2) 48kg X m/s
(3) 20kg X m/s   ✓ (4) 0 kg X m/s

4. A bullet of mass 0.05kg leaves the muzzle of a gun of mass 4.0kg with a velocity of 800m/s. What is the velocity of the recoil of the gun?

5. If there was some friction between an ice skater's skates and the ice, would momentum be conserved?

6. Momentum is equal to
(1) force times time   (2) mass times acceleration
(3) acceleration times time ✓(4) none of these

7. A 40kg girl is standing on a stationary 10 kg wagon and jumps off the back of the wagon, giving the wagon a kick and sends it off in the opposite direction at 4m/s. How fast is the girl moving (fig.17)?

fig. 17

8. The same 40kg girl is now runnimg along at 5m/s and jumps onto the stationary 10kg wagon and they move together in the same direction as she

fig. 18

was running. How fast is the wagon
moving with the girl in it (fig.18)?

9. A 4kg basketball traveling at 5m/s
hits the back of a stationary 10kg
wagon and bounces off at 4m/s in the
path it came from and sends the wagon
off in the original direction of the
ball. How fast is the wagon moving
(fig.19)?

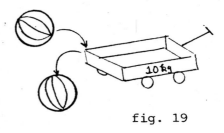

fig. 19

$$(4kg)(5m/s) = (4kg)(-4m/s) + 10kg(V)$$
$$20 kg·m/s = -16 kg·m/s + 10 kg·V$$
$$3.6 m/s$$

MATH HUMOR VI

1. Prove: II = VI by moving one line

2. Prove: 6+5=9

3. Why did the physicist disconnect his door bell?

4. Prove: sin x = 6n

5. In how many ways can you get forty cents?

HA!
HA!

Chapter VII

MIDTERM REVIEW

1. **ACCELERATION:**
   THE RATE CHANGE IN VELOCITY WITH RESPECT TO TIME. IT
   MAY BE A CHANGE IN SPEED, IN DIRECTION, OR IN BOTH SPEED
   AND DIRECTION.

$$\text{ACCELERATION} = \frac{\text{CHANGE IN VELOCITY}}{\text{TIME}}$$

QUESTIONS:

1-1. An object whose acceleration is zero may be
   - (1) starting
   - (2) stopping
   - (3) changing direction
   - ✓ (4) moving at a constant speed

1-2. Acceleration is a vector quantity that represents the
   time-rate of change in
   - (1) momentum (2) velocity ✓ (3) distance (4) energy

1-3. Which statement about the movement of an object
   with zero acceleration is true?
   - (1) The object must be at rest.
   - (2) The object must be slowing down.
   - (3) The object may be speeding up.
   - ✓ (4) The object may be in motion.

1-4. The unit "miles per hour per second" measures
   - (1) velocity (2) direction (3) speed (4) acceleration ✓

1-5. Acceleration can be expressed as
   - (1) m/s (2) m^2/s (3) m^2/s^2 ✓   (4) m/s^2

2. **ACCELERATION DUE TO GRAVITY:**
   IT IS THE ACCELERATION OF A FREELY FALLING OBJECT AND
   IS USUALLY REPRESENTED BY THE LETTER "g".
   THE VALUE OF "g" AT MOST POINTS NEAR THE EARTH'S
   SURFACE IS

$$g = 9.8 m/s^2$$

QUESTIONS:

2-1. If a freely falling object were monitored, it would register an increase in speed each second by

    (1) 4.9 m/s   ✓(3) 9.8m/s

    (2) $9.8m/s^2$ (4) depends on the initial speed

2-2. As an object freely falls its

    ✓(1) velocity increases   (3) both (1) and(2)

    (2) acceleration increases   (4) neither (1) nor (2)

2-3. If an object falls with constant acceleration, the velocity of the object must

    (1) also be constant

    ✓(2) be continuously changing by the same amount each second

    (3) be continuously changing by different amounts depending on the speed

    (4) be continuously decreasing

2-4. Which is constant for a freely falling object?

    (1) displacement   (3) velocity

    (2) speed   ✓(4) acceleration

2-5. A freely falling object near the Earth's surface has a constant

    (1) velocity of 1.0m/s   (3) acceleration of $1.0m/s^2$

    (2) velocity of 9.8m/s   ✓(4) acceleration of $9.8m/s^2$

3. **ANGULAR MOMENTUM:**

A QUANTITY MEASURING THE ROTATION OF AN OBJECT ABOUT ITS AXIS OR THE REVOLUTION OF AN OBJECT ABOUT AN EXTERNAL AXIS. FOR AN OBJECT HAVING A GIVEN MASS, MOVING AT A GIVEN SPEED IN A CIRCULAR PATH AT A GIVEN DISTANCE FROM A SECOND OBJECT THE MAGNITUDE OF THE ANGULAR MOMENTUM IS 'MASS X SPEED X RADIUS' OR 'MVR'. HOWEVER, THE ANGULAR MOMENTUM OF A ROTATING OBJECT IS CALCULATED ON THE CONCEPT THAT THE OBJECT HAS MANY ELEMENTS AND EACH ELEMENT HAS ITS OWN ANGULAR MOMENTUM. WHEN ADDED TOGETHER, THIS PRODUCES ANOTHER FORMULA FOR THE TOTAL ANGULAR MOMENTUM OF A ROTATING OBJECT.

4. __CENTER OF GRAVITY__:
    THE AVERAGE POSITION OF THE WEIGHT OF AN OBJECT. IT
    IS THE POINT AT WHICH AN OBJECT WILL NOT ROTATE IF IT
    IS SUPPORTED THERE.

5. __CENTER OF MASS__:
    THE AVERAGE LOCATION OF THE MASS OF AN OBJECT.

6. __CENTRIFUGAL FORCE__:
    A "CENTER-FLEEING" FORCE WHICH IS A FORCE DIRECTED AWAY
    FROM THE CENTER OF A CIRCLE. IT APPEARS THAT THE FORCE
    IS ALONG THE RADIUS IN AN OUTWARD DIRECTION. ALTHOUGH
    THIS IS REFERRED TO AS A "FICTITIOUS FORCE", THE TERM
    APTLY DESCRIBES THE FORCE THAT AN INDIVIDUAL THINKS
    HE OR SHE IS EXPERIENCING.

7. __CENTRIPETAL FORCE__;
    A "CENTER-SEEKING" FORCE, WHICH IS A FORCE DIRECTED
    TOWARD THE CENTER OF A CIRCLE. IT CAUSES AN OBJECT TO
    ACCELERATE INTO THE CENTER AND WOULD RESULT IN THE
    OBJECT MOVING IN A CIRCULAR PATH RATHER THAN A STRAIGHT
    LINE.

__QUESTIONS__:

7-1. The force of _____ is the centripetal force acting on
      the Moon.
      (1) motion (2) gravity (3) air resistance (4) the Sun

7-2. If the force that keeps an object moving in a circle
      is removed, the object will
      (1) stop   (3) move in a straight line
      (2) reverse directions   (4) continue to move in a circle

7-3. The centripetal force acting on the
      airplane in the diagram at the
      right is directed toward point
      (1) A   (2) B   (3) C   (4) D

7-4. A ball is attached to a string and is swung in a circle in a horizontal plane. The force keeping the ball moving in a circle is
   (1) tangent to the circle
   (2) perpendicular to the radius of the circle
   (3) perpendicular to the tangent to the circle and directed away from the center of the circle
   √ (4) perpendicular to the tangent to the circle and directed toward the center of the circle

7-5. The centripetal acceleration of a satellite is approximately
   (1) 9.8m/s   √(3) 4.9m/s
   √ (2) 9.8m/s^2 (4) 4.9m/s^2

7-6. The force required to keep an object moving in a circle is called the
   (1) centrifugal force   (3) tangential force
   √(2) centripetal force   (4) linear force

7-7. A satellite is held in orbit around the Earth because of
   (1) centrifugal force   √(3) the Earth's gravity
   (2) the Sun's gravity   (4) the Moon's gravity

8. **COMPONENT:**
   THE VECTOR PARTS INTO WHICH A GIVEN VECTOR MAY BE RESOLVED.

   QUESTIONS:

8-1. If the force vector shown in the diagram below is resolved into two components, these two components could be represented by which diagram?

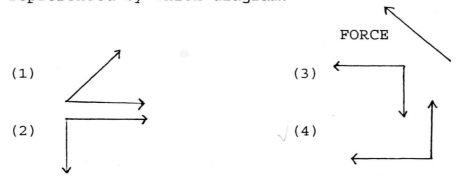

-74-

8-2. The diagram at the right represents the components into which a given vector has been resolved. Which of the following choices could that vector be?

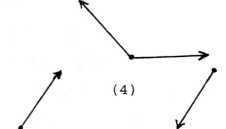

(1) ←———•   (2) •↘   ✓(3)   (4)

9. **CONSERVATION OF ANGULAR MOMENTUM:**
IN THE ABSENCE OF AN UNBALANCED TORQUE THE TOTAL ANGULAR MOMENTUM BEFORE IS EQUAL TO THE TOTAL ANGULAR MOMENTUM AFTER.

$$\text{mvr}_{before} \quad = \quad \text{mvr}_{after}$$

10. **CONSERVATION OF MOMENTUM:**
THE TOTAL MOMENTUM OF A SYSTEM WITH ONLY INTERNAL FORCES (NO OUTSIDE FORCES) BEFORE AN INTERACTION IS EQUAL TO THE TOTAL MOMENTUM AFTER THE INTERACTION.

$$\text{THE TOTAL MOMENTUM BEFORE} \quad = \quad \text{THE TOTAL MOMENTUM AFTER}$$

$$\text{mv}_{before} \quad = \quad \text{mv}_{after}$$

QUESTION;
10-1. The conservation of momentum is most closely related to
    (1) Newton's First Law   ✓(3) Newton's Third Law
    (2) Newton's Second Law   (4) not enough information

11. **ESCAPE VELOCITY:**
THE MINIMUM VELOCITY REQUIRED TO ESCAPE THE GRAVITATIONAL PULL OF ANOTHER CELESTIAL BODY. IT IS THE VELOCITY REQUIRED BY A SPACE PROBE TO OVERCOME THE EARTH'S GRAVITATIONAL FORCE.

12. **FORCE:**
A PUSH OR A PULL CAUSING AN OBJECT TO ACCELERATE. IT IS MEASURED IN NEWTONS.

$$1N = 1 kg \cdot m/s^{2}.$$

QUESTIONS:

12-1. The S.I. unit of force is the _____?
 (1) kg   (2) meter   (3) cm   √(4) Newton

12-2. Two forces are applied to a 5.0kg block on a frictionless horizontal surface as shown in the diagram.

$$3.0N \rightarrow \boxed{5.0kg} \leftarrow 7.0N$$

 The net force is
 (1) 4.0N to the right   (3) 10.0N to the right
 √(2) 4.0N to the left   (4) 10.0N to the left

13. **FREE-FALL**:
 WHEN AN OBJECT FALLS TO EARTH WITH ONLY GRAVITY ACTING ON IT.

QUESTIONS:

13-1. In free-fall which of the following quantities increases with time?
 (1) acceleration   √(3) velocity
 (2) mass   (4) none of these

13-2. If a ball is rolled off the edge of a table, its horizontal component of motion
 (1) increases   √(3) remains the same
 (2) decreases   (4) not enough information

13-3. As the mass of a freely falling object increases, the acceleration with which it falls
 (1) increases   √(3) remains the same
 (2) decreases   (4) increases then decreases

13-4. A ball thrown by a player on a team
 (1) has a falling speed that is directly proportional to its horizontal speed
 (2) only begins to fall when its horizontal speed equals zero
 √(3) has a falling speed that is independent of its horizontal speed
 (4) not enough information

13-5. If two objects of different masses are released at the same time from an elevated position above the floor then, ignoring air resistance,

    (1) the lighter mass reaches the floor first
    (2) the heavier mass reaches the floor first
  √(3) they both reach the floor at the same time
    (4) not enough information

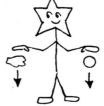

14.   **FRICTION**:

    A FORCE THAT OPPOSES MOTION.

15.   **INERTIA**:

    THE PROPERTY WHEREBY AN OBJECT OPPOSES A CHANGE IN MOTION, THAT IS, IT REMAINS AT REST OR IT CONTINUES TO MOVE IN A STRAIGHT LINE UNLESS IT IS ACTED UPON BY AN EXTERNAL FORCE.

    QUESTIONS:

15-1. A resistance to having the state of motion changed is called

    (1) gravity   (2) friction   (3) force   (4) inertia

15-2. Compared to the inertia of a 0.20kg ball, the inertia of a 0.40kg ball is

    (1) one-half as great   (3) one-fourth as great
    (2) twice as great   (4) four times as great

15-3. According to Galileo

    (1) in the absence of air resistance heavier objects fall faster than lighter objects
    (2) an object in motion continues in motion in a straight line unless an external force is imposed upon it
    (3) an object belongs at rest unless a violent force is imposed upon it
    (4) the Earth is the center of the universe

16.   **IMPULSE**:

    A CHANGE IN MOMENTUM EITHER BY STOPPING AN OBJECT IN MOTION OR BY IMPARTING MOTION TO AN OBJECT. IT IS EQUAL TO THE FORCE ACTING ON THE OBJECT TIMES THE LENGTH OF TIME THE FORCE ACTS.

$$I = F \times T$$

QUESTIONS:

16-1. It is correct to say that impulse is equal to
(1) momentum
(2) the change in momentum
(3) the force multiplied by the distance the force acts
(4) none of the above

16-2. A force of 10N acts on an object for 0.010 seconds.
What force, acting on the object for 0.050 seconds, would
produce the same results?
(1) 1N   (2) 2N   (3) 5.0N   (4) 10N

16-3. Boxing gloves reduce injury by
(1) increasing style   (3) increasing contact time
(2) a quicker blow   (4) not enough information

17.   **KEPLER'S LAWS OF PLANETARY MOTION.**
LAW I: EACH PLANET MOVES IN AN ELLIPTICAL ORBIT WITH
THE SUN AT ONE FOCUS.
LAW II: THE LINE FROM THE SUN TO ANY PLANET SWEEPS OUT
EQUAL AREAS OF SPACE IN EQUAL TIME INTERVALS.
LAW III: THE SQUARES OF THE PERIODS OF REVOLUTION OF
THE PLANETS ARE PROPORTIONAL TO THE CUBES OF
THEIR AVERAGE DISTANCES FROM THE SUN.
$$R^3 / T^2 = \text{constant}$$

QUESTIONS:

17-1. According to Kepler's Laws, the paths of planets about
the Sun are
(1) parabolas (2) circles (3) straight lines (4) ellipses

17-2. An Earth satellite in elliptical orbit travels fastest
when it is
(1) nearest the Earth
(2) farthest from the Earth
(3) neither far or near because it travels at a constant
speed all along the orbit
(4) not enough information

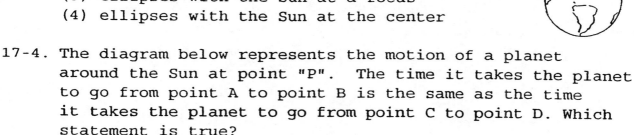

17-3. The shapes of the paths of the planets about the Sun are all
    (1) circles with the Sun at the center
    (2) circles with the Sun off center
    (3) ellipses with the Sun at a focus
    (4) ellipses with the Sun at the center

17-4. The diagram below represents the motion of a planet around the Sun at point "P". The time it takes the planet to go from point A to point B is the same as the time it takes the planet to go from point C to point D. Which statement is true?

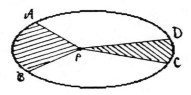

    (1) The two shaded areas are equal in area.
    (2) As the planet orbits the Sun, it moves at a constant speed
    (3) The planet moves faster from point C to point D because it is farther from the Sun.
    (4) not enough information.

17-5. Which of the following statements is credited to Kepler?
    (1) He developed a theory of gravity that could explain orbital motion.
    (2) He found with the use of a telescope that the Earth is the center of the universe.
    (3) He found that the planets move around the Sun in elliptical orbits.
    (4) He found that the planets move around the Sun in circular orbits.

18. **KILOGRAM**:
    STANDARD OF MASS IN THE METRIC SYSTEM.

19. **MASS**:
    A MEASURE OF AN OBJECT'S INERTIA WHICH IS ITS RESISTANCE TO CHANGE. IT IS ALSO REFERRED TO AS A QUANTITY OF MATTER.

19-1. As a satellite is accelerated away from the Earth by
a rocket, the satellite's mass
(1) increases   (3) remains the same
(2) decreases   (4) not enough information

19-2. Compared to the mass of a 10N object on the Earth, the
mass of the same object on the Moon is
(1) the same   (3) less
(2) greater   (4) not enough information

20.   **METER:**
STANDARD OF LENGTH IN THE METRIC SYSTEM.

QUESTIONS:
20-1. Which term represents a fundamental unit?
(1) Watt   (2) Newton   (3) meter   (4) Joule

20-2. The metric system is easy to use because it is based
on multiples of the number_____?
(1) 12   (2) 20   (3) 10   (4) 11

20-3. Which measurement of an average door is closest to 1m?
(1) the thickness   (3) the height
(2) the surface area   (4) the height from the floor
                                     to the door knob

21.   **M.K.S. SYSTEM:**
**A PART OF THE S.I. SYSTEM. THE LETTERS REFER TO THE METER,
THE KILOGRAM, AND THE SECOND, RESPECTIVELY, AS STANDARD
FUNDAMENTAL UNITS.**

QUESTION:
21-1. The unit of length in the metric system is the
(1) kilogram   (2) foot   (3) meter   (4) inch

22.   **MOMENTUM:**
THE PRODUCT OF THE MASS AND VELOCITY OF A MOVING OBJECT.

QUESTIONS:
22-1. The direction of an object's momentum is always the same
as the direction of the object's
(1) inertia (2) velocity (3) potential energy (4) weight

22-2. A train rolls along a track with considerable momentum. If it rolls at the same speed but has twice as much mass, its momentum is
(1) zero   (2) doubled   (3) quadrupled   (4) unchanged

22-3. Momentum may be expressed in
(1) Joules   (2) Watts   (3) kg·m/s   (4) Newtons

22-4. As an object falls freely toward the Earth its momentum
(1) decreases   (3) remains the same
(2) increases   (4) not enough information

22-5. The momentum of an object is the product of its
(1) mass and acceleration   (3) force and displacement
(2) mass and velocity   (4) not enough information

22-6. A rocket with a mass of 1000kg is moving at a speed of 20m/s. The magnitude of the momentum is
(1) 50 kg·m/s   (3) 20,000 kg·m/s
(2) 2000 kg·m/s   (4) 400,000 kg·m/s

22-7. A heavy truck has more momentum than a passenger car moving at the same speed because the truck
(1) is not streamlined   (3) has greater velocity
(2) has a larger wheelbase   (4) has greater mass

22-8. The greater the velocity of an object the greater is its
(1) momentum   (3) weight
(2) potential energy   (4) mass

22-9. Two rocks weighing 10N and 20N fall freely from rest near the Earth's surface. If air resistance is ignored, which of the following statements is true?
(1) The 10N rock will have a greater acceleration than the 20N rock.
(2) The 20N rock will have a greater acceleration than the 10N rock.
(3) At the end of 1 second of falling freely, the 20N rock will have a greater momentum than the 10N rock.
(4) At the end of 1 second of falling freely, the 10N rock will have a greater momentum than the 20N rock.

23. **NEWTON:**
ONE NEWTON OF FORCE IS REQUIRED TO ACCELERATE A MASS OF
ONE KILOGRAM AT A RATE OF ONE METER PER SECOND PER SECOND.
$$1N = 1 \text{ kg} \cdot \text{m/s}^2$$

QUESTIONS:

23-1. Which of the following is a derived unit?
   (1) second   (2) meter   (3) kilogram   (4) Newton

23-2. The fundamental units for a force of one Newton are
   (1) meters/second   (3) meters/second/kilogram
   (2) kilograms   (4) kilograms·meters/second^2

24. **NEWTON'S LAWS OF MOTION:**

   LAW I: EVERY OBJECT CONTINUES IN ITS STATE OF REST,
                OR OF UNIFORM MOTION IN A STRAIGHT LINE, UNLESS
                IT IS COMPELLED TO CHANGE THAT STATE BY FORCES
                IMPRESSED UPON IT. THIS IS KNOWN AS THE LAW
                OF INERTIA.

   LAW II: IF AN UNBALANCED FORCE IS IMPOSED ON AN OBJECT,
                IT WILL ACCELERATE. THE ACCELERATION IS DIRECTLY
                PROPORTIONAL TO THE NET FORCE, IS IN THE SAME
                DIRECTION AS THE NET FORCE AND IS INVERSELY
                PROPORTIONAL TO THE MASS OF THE OBJECT.

$$\text{ACCELERATION} = \frac{\text{FORCE}}{\text{MASS}}$$

   or FORCE = MASS x ACCELERATION

   LAW III: IF ONE OBJECT IMPOSES A FORCE ON A SECOND OBJECT,
                THE SECOND OBJECT WILL IMPOSE AND EQUAL AND
                OPPOSITE FORCE ON THE FIRST OBJECT. THIS IS
                ALSO KNOWN AS THE LAW OF ACTION AND REACTION.

QUESTIONS:

24-1. In Newton's Third Law, the reaction force is
   (1) greater than the action force
   (2) less than the action force
   (3) in the same direction as the action force
   (4) equal to the action force

24-2. An object weighing 4N rests on a horizontal tabletop.
The force of the table on the object is
(1) 0 N (2) 4N horizontally (3) 4N downward (4) 4N upward

24-3. Two equal and opposite forces of 7N have a net force of
(1) 10N   (2) 7N   (3) 3½N   (4) zero N

24-4. Which of the following statements is accurate about why
a book is resting on a table?
(1) There is a net force acting downward on the book.
(2) There is a net force acting upward on the book.
(3) The weight of the book always equals the weight of
the table.
(4) The weight of the book and the table's upward force
on the book are equal in magnitude, but opposite
in direction.

24-5. Which could be expressed in units of mass only?
✓(1) force/acceleration   (3) momentum/time
(2) power X time   (4) energy/force

24-6. Two equal forces are acting in opposite directions. If
one force increases in magnitude while the other force
decreases in magnitude then the resultant force
(1) decreases   (3) remains the same
✓(2) increases   (4) not enough information

24-7. Newton's _____ Law of Motion explains how a jet or
rocket engine works.
(1) 1st   (2) 2nd   (3) 3rd   (4) not enough information

24-8. If equal forces are applied to different masses, the
greater mass will experience the _____ acceleration.
(1) greater (2) smaller (3) same (4) not enough information

24-9. Newton's First Law of Motion is a statement of the law
of
(1) inertia   (3) net force
(2) action-reaction   (4) universal gravitation

24-10. In Newton's Third Law of Motion the action and reaction forces
   (1) act on the same object
   (2) act on different objects
   (3) are equal and are in the same direction
   (4) are unequal and are in opposite directions

25. **PARABOLA**:
   A CURVED LINE REPRESENTING THE PATH OF A PROJECTILE.

26. **PROJECTILE**:
   ANYTHING THAT IS THROWN, SUCH AS A ROCK, OR ANYTHING THAT IS SHOT, SUCH AS FROM A RIFLE OR FROM A CANNON.

27. **RESULTANT**:
   A FORCE REPRESENTING THE MATHEMATICAL SUM OF TWO OR MORE FORCES ACTING AT THE SAME TIME.

QUESTIONS:

27-1. The single force that has the same effect as two forces which it replaces is known as their
   (1) replica   (2) equilibrium   (3) resultant   (4) vector

27-2. Three forces act on an object: 5N, 8N, and 12N. The object remains stationary and no other forces act on it. The resultant of all three forces is
   (1) 0 N   (2) 25N   (3) 21N   (4) cannot be determined since no direction is given

27-3. The resultant of a 12N force and a 5N force acting simultaneously on an object in the same direction is
   (1) 0 N   (2) 5N   (3) 7N   (4) 17N

27-4. Which vector represents the resultant of the two vectors shown?

27-5. Which force could act concurrently with force A to produce force B as resultant?

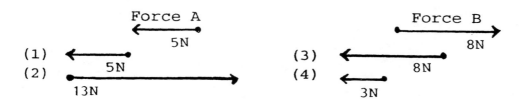

Force A
5N

Force B
8N

(1) 5N

(3) 8N

(2) 13N

(4) 3N

28.　**REVOLUTION**:
CIRCULAR MOTION OF AN OBJECT AROUND ANOTHER OBJECT NOT ATTACHED TO THE FIRST OBJECT.

29.　**ROTATION**:
TO SPIN AROUND ON AN IMAGINARY AXIS WITHIN THE SPINNING OBJECT.

30.　**ROTATIONAL INERTIA**:
RESISTANCE TO CHANGE IN ROTATIONAL MOTION.　IT DEPENDS ON THE MASS OF THE OBJECT AND THE MASS DISTRIBUTION ABOUT THE AXIS OF ROTATION.

31.　**SCALAR QUANTITY**:
A QUANTITY THAT DESCRIBES MAGNITUDE OR SIZE.　IT DOES NOT SHOW DIRECTION.

QUESTION:
31-1. Which is a scalar quantity?
(1) weight　　　　　(3) velocity
(2) mass　　　　　 (4) force

32.　**SECOND**:
STANDARD UNIT OF TIME IN THE METRIC SYSTEM.

33.　**S.I.**:
SYSTEMS INTERNATIONALE.　STANDARD FUNDAMENTAL UNITS FORMING THE UNIFORM MEASUREMENTS OF THE METRIC SYSTEM.

34. **TANGENT:**
    A STRAIGHT LINE TOUCHING A CURVE IN ONE AND ONLY ONE POINT NO MATTER HOW FAR THE STRAIGHT LINE IS EXTENDED IN EITHER DIRECTION.

QUESTION:

34-1. Which diagram represents the vector for the tangential velocity of a satellite.

(1) (2) (3) (4)

35. **TERMINAL VELOCITY:**
    THE GREATEST VELOCITY REACHED BY A FALLING OBJECT.

QUESTION:

35-1. The speed a falling object reaches when the downward acceleration of gravity is just balanced by the upward effect of air resistance is called_____.
    (1) acceleration   (3) weight
    (2) motion   (4) terminal velocity

36. **TORQUE:**
    **PRODUCES A CHANGE IN ROTATIONAL MOTION.
    IT IS EQUAL TO THE PRODUCT OF THE LEVER
    ARM TIMES THE APPLIED FORCE.**

QUESTIONS:

36-1. If two people are balanced on a see-saw and one person moves inward toward the pivot point, then that person's side will
    (1) rise   (3) neither rise nor fall
    (2) fall   (4) not enough information

36-2. Radius times tangential force equals_____.
    (1) weight   (2) mass   (3) torque   (4) freefall

36-3. A torque acting on an object tends to produce
    (1) equilibrium   (3) a center of gravity
    (2) velocity   (4) rotation

36-4. Michael and David are balanced on a see-saw. They have equal
   (1) distances from the fulcrum   (3) masses
   (2) weights   (4) torques

37.   **LAW OF UNIVERSAL GRAVITATION:**
   EVERY OBJECT IN THE UNIVERSE ATTRACTS EVERY OTHER OBJECT WITH A FORCE THAT IS DIRECTLY PROPORTIONAL TO THE PRODUCT OF THEIR MASSES AND IS INVERSELY PROPORTIONAL TO THE SQUARE OF THE DISTANCE BETWEEN THEIR CENTERS OF MASS.

$$F = G\frac{m_1 m_2}{d^2}$$

G = UNIVERSAL CONSTANT
G = 6.67 X 10^{-11} N·m²/kg²

QUESTIONS:

37-1. As two objects that are moving toward each other by mutual gravitation get closer and closer, the force between them
   (1) increases
   (2) decreases
   (3) remains the same
   (4) not enough information

37-2. The constant "G" is
   (1) the acceleration due to gravity
   (2) smaller on the Moon than on the Earth
   (3) the force of gravity
   (4) a universal constant

37-3. According to Newton, the greater the masses of intersecting objects the
   (1) less the gravitational field between them
   √(2) greater the gravitational field between them
   (3) greater the force between them by the square of the masses
   (4) none of the above

37-4. Two objects of equal mass are a fixed distance apart. If the mass of each object could be tripled, the gravitational force between the objects would
(1) decrease by one-third
(2) decrease by one-ninth
(3) triple
(4) increase nine times

37-5. Two point masses that are equal are separated by a distance of one meter. If one mass is doubled the gravitational force between the two masses would be
(1) ½ as great   (3) ¼ as great
✓(2) two times as great   (4) four times as great

37-6. As the mass of an object increases, its' gravitational force of attraction on the Earth
✓(1) increases   (3) remains the same
(2) decreases   (4) not enough information

37-7. If the Moon were twice as massive as it is now, then the gravitational force between it and the Earth would be
(1) ½ as great   (3) ¼ as great
✓(2) two times as great   (4) the same

37-8. The force of gravity between two objects
(1) increases with the increasing masses of the objects and decreases with the increasing distance between them
(2) decreases with the decreasing masses of the objects and increases with the increasing distance between them
(3) increases with the decreasing masses of the objects and increases with the decreasing distance between them
(4) decreases with the increasing masses of the objects and decreases with the decreasing distance between them

37-9. Compared to the mass of a 10N object on the surface of the Earth, the mass of the same object a distance of three Earth radii from the center of the Earth is
(1) the same   (3) ½ as great
(2) two times as great   (4) ¼ as great

38. **VECTOR:**
A QUANTITY THAT HAS MAGNITUDE WHICH IS A NUMERICAL AMOUNT AND ALSO HAS DIRECTION. IT IS USUALLY REPRESENTED BY AN ARROW WITH THE LENGTH PROPORTIONAL TO THE MAGNITUDE AND THE ARROWHEAD REPRESENTING THE DIRECTION.

QUESTIONS:

38-1. Which term represents a vector quantity?
(1) distance   (2) speed   (3) temperature   (4) force

38-2. Which pair of terms are vector quantities?
(1) force and mass   (3) distance and displacement
(2) speed and velocity   (4) momentum and acceleration

38-3. A person walks 5 blocks north, 5 blocks west and 5 blocks south. What is the displacement of the person?
(1) 5 blocks east   (3) 5 blocks north
(2) 5 blocks west   (4) 5 blocks south

38-4. A car travels 30m east in 4.0 seconds. The displacement of the car at the end of the 4.0 second interval is
(1) 30m   (2) 30m/s   (3) 30m east   (4) 30m/s east

38-5. Which is a vector quantity?
(1) the temperature on the coldest day of the year
(2) the number of people in a room
(3) the number of test scores over ninety
(4) the weight of an object

38-6. A quantity that has both size and direction is a
(1) scalar   (2) vector   (3) kilogram   (4) meter

38-7. Which is not a vector?
(1) speed   (2) velocity   (3) resultant   (4) momentum

38-8. A baseball player runs 90' from home to first base, then over runs it by 5' and returns to first base. What is the distance the player has run?
(1) 85'   (2) 90'   (3) 95'   (4) 100'

38-9. Using the same information in question 38-8, what is the magnitude of the players displacement from home base?
(1) 85'   (2) 90'   (3) 95'   (4) 100'

39. **VELOCITY**:
THE SPEED OF AN OBJECT AND ITS DIRECTION OF MOTION. IT
IS A VECTOR QUANTITY.

QUESTION:
39-1. An object whose velocity is zero is
    (1) at rest   (3) accelerating
    (2) in motion   (4) changing direction

40. **WEIGHT**:
THE FORCE WITH WHICH AN OBJECT IS ATTRACTED TO THE EARTH
       WEIGHT = MASS X ACCELERATION DUE TO GRAVITY
         W = mg

41. **WEIGHTLESSNESS**:
HAVING NO APPARENT WEIGHT.

QUESTION:
41-1. There is no perceptible gravity within an orbiting
spacecraft because the spacecraft is in
    (1) free-fall (2) lift off (3) reverse (4) the dark

MATH HUMOR VII

1. What is the next letter in this sequence O,T,T,F,F,S,S,...?

2. Circle the numbers that add to 21.

       9 9 9
       5 5 5
       3 3 3
       1 1 1

3. A young man went away to college to get a B.A. in
Mathematics. Since his father paid the tuition, he asked
his father to also pay for his Masters in Mathematics.
His father agreed. After receiving his Masters, he then
asked his father to pay the tuition for his PhD. in
Mathematics and again his father agreed. After getting
his PhD., he came home to celebrate. His mother made dinner

for the three of them. Just before dinner his father asked him to explain some of the mathematics he learned. He said, "You would not understand the new math. Do you see these two chickens on the table? There are really three chickens." The father, in disgust, said, "Let's eat." The father took one of the chickens for himself, gave the other to his wife and said, "You, my educated son, can have the third chicken."

4. What did one math tree say to the other math tree?

5. A TEST TAKERS nightmare: Taking a 100 question multiple choice test and having all the answers come out as "A".

Chapter VIII

PRACTICE MIDTERM QUESTIONS

1. The unit of length in the metric system is the
    (1) kilogram   (2) foot   (3) meter   (4) light year
2. The S.I. unit of velocity is the
    (1) foot per second   (3) yard per hour
    (2) meter per second   (4) light year
3. The rate of change of velocity is called
    (1) change in speed   (3) change in direction
    (2) slowing down   (4) acceleration
4. The ability of matter to resist changes in its state of motion is
    (1) mass   (3) inertia
    (2) weight   (4) momentum
5. All objects in free fall in a given place have the same
    (1) acceleration (2) size   (3) area   (4) weight
6. According to Newton's First Law, every moving object has
    (1) motion (2) acceleration (3) inertia   (4) height
7. According to Newton's Third Law, when a force is exerted on an object, the object exerts an equal force in the
    _____ direction.
    (1) same   (2) combined   (3) opposite   (4) overall
8. If a car is going northward and starts to turn right, the direction of its acceleration is
    (1) north   (2) west   (3) south   (4) east
9. When an object is in orbit, its acceleration is produced by
    (1) friction   (2) inertia (3) gravity (4) centrifugal force
10. The amount of material in an object is called the object's
    (1) mass   (2) height   (3) weight   (4) width
11. If Melanie stands next to a wall and pushs it with a force of 20N, it will push her with a force of _____ N.
    (1) 5   (2) 10   (3) 15   (4) 20
12. The force of gravity between two objects is directly proportional to the product of their
    (1) weight (2) inertia   (3) momentum   (4) masses
13. If an objects' momentum is directed to the west, then the objects'_____ is also directed to the west.
    (1) velocity (2) mass   (3) impulse   (4) torque

14. The centripetal force acting on a satellite is produced by
    (1) friction (2) inertia   (3) air resistance (4) gravity
15. A force applied to an object may
    (1) start it   (3) change its direction
    (2) stop it   (4) all of these
16. In Newton's Third Law, the action force must be
    (1) greater than the reaction force
    (2) less than the reaction force
    (3) in the same direction as the reaction force
    (4) none of the above
17. A person traveling from the Earth to the Moon will find
    (1) his weight and his mass are both less.
    (2) his weight and his mass are both greater.
    (3) his weight is the same and his mass is less.
    (4) his mass is the same and his weight is less.
18. A scientist who described the concepts in Newton's Laws
    differently when dealing with motion at extremely high speeds
    was
    (1) Aristotle (2) Einstein (3) Copernicus (4) Kepler
19. Which two quantities are measured in the same units?
    (1) mass and weight   (3) velocity and acceleration
    (2) weight and force   (4) force and momentum
20.   A 4.0 kg mass is moving in a circular path
                              at the end of a string and moves in a
                              horizontal plane at a constant speed.
                              The centripetal force acting on the mass
                              is directed toward
                              (1) A   (2) B   (3) C   (4) D
21. Two objects of equal mass are a fixed distance apart. If
    the mass of each object could be doubled, the gravitational
    force between the objects would
    (1) decrease by one-half   (3) double
    (2) decrease by one-fourth (4) increase four times
22. Momentum is the product of mass and
    (1) velocity (2) time   (3) force   (4) acceleration
23. The velocity of a falling body at the point where air
    resistance just balances the gravitational force is called
    (1) acceleration   (3) impact
    (2) terminal velocity   (4) momentum
24. In an isolated system the total angular momentum is
    (1) conserved (2) decreased (3) increased (4) zero

25. Angular momentum of a point mass increases, with no change in speed, when the object is moved _____ its center of revolution.

(1) towards   (2) passed   (3) away from   (4) over

MATH HUMOR VIII

1. Add, subtract, multiply or divide eight eights to get an answer of 1000.

2. Add two matches to three matches to make eight.

3. Pick any two numbers. For example, 65 and 28. Add these two numbers. Add your answer to the second number. Now add your new answer to the previous answer. Repeat this procedure until you have ten numbers from the beginning. Add all ten numbers. How could you get the sum without doing thie final addition?

4. What is the next symbol in the following sequence?

M ♡ 8 ⚛ ?

5. Given a stack of eight dice as shown in the diagram below. Find the sum of the numbers on the hidden surfaces.

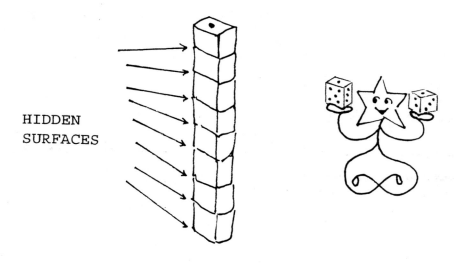

HIDDEN
SURFACES

𝔆𝔥𝔞𝔭𝔱𝔢𝔯 IX

WORK AND ENERGY

The Universe is made up of matter and energy. Matter has mass, occupies space and can be seen, felt, tasted and smelled. Energy, on the other hand, is a concept and has none of the properties of matter.

impulse = force × time

In Chapter VI the idea of impulse was discussed and shown to be equal to force times time. Impulse which is a change in momentum depends upon a force and the length of time the force acts on the object. Now we are going to consider the relationship of an applied force and the distance through which the force acts in the direction of the force. In physics this is called work.

WORK = FORCE X DISTANCE MOVED IN THE DIRECTION OF THE FORCE
W = fd

If we push on a wall with all our force and the wall does not move, then we did not do any work. According to the physicist, work equals force times distance and the distance in this situation equals zero. Any quantity times zero equals zero. This means motion in the direction of the force must be present for any effort to be called work. If a person is just holding an object which is very heavy and is getting tired, the person is not doing any work. If the object that the person is holding did not move, then no work was done. One might conclude that no work is done only if an object is not in motion. This is not the case. Even if an object is in motion there may be no work involved. For example, no work is done if an object is moving with a constant velocity. Remember Newton's First Law which states that an object in motion will remain in motion in a straight line unless an external force is imposed on it. A constant velocity implies no change in speed or direction. Since the velocity is constant, no force is being applied to change its

fig. 1

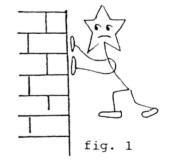

fig.2

speed or direction and it will continue to move in a straight line indefinitely. Work requires a force times the distance in the direction of the force. If there is no force, there is no work involved. In addition, there is no work done if an object moves with a constant speed in a circular path. Notice the use of the words, "constant speed" instead of "constant velocity". In the case of uniform circular movement, the speed does not change but the direction is changing which means we cannot say constant velocity. When the object moves in a circular path the force is applied along the radius perpendicular to the tangential path of the object. Without a force in the direction of the motion, there is no work done. An example of this would be a satellite orbiting the Earth. It does not necessarily follow that if there is motion, there is work done.

Since force is measured in newtons and distance is measured in meters, then work is measured in newtons and meters which together is referred to as joules. One joule of work is done when a force of one newton is exerted for a distance of one meter. Joule is pronounced like pool and is named after the scientist James Prescott Joule who was the first to calculate the mechanical equivalent of heat. In summary, if we push an object with a force of 2 newtons in the direction of motion for a distance of 10 meters we have done 20 joules of work,

$$W = fd$$
$$= 2N \times 10m$$
$$= 20J$$

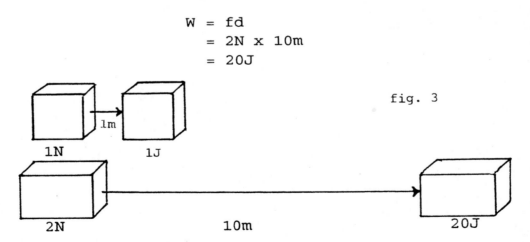

fig. 3

Sometimes a force acts at an angle to the motion. In such cases, it is the component of the force along the direction of motion that enters into the work calculation.

	UNIT OF FORCE	MEASURE OF WORK
MKS SYSTEM	Newton (N)	N•m = Joule
BRITISH SYSTEM	Pound (lb.)	ft•lb
CGS SYSTEM	Dyne	dyne•cm = erg

In the diagram at the right (fig. 4) a box is being pushed up a hill from point A to point B. The output work done in elevating the box to point B is equal to the weight force (1000N) times the height through which the box is raised (2.0m). The output work in getting the box from point A' to point C is the same because the weight of the box is the same and the height of point C is the same as the height of point B.

fig. 4

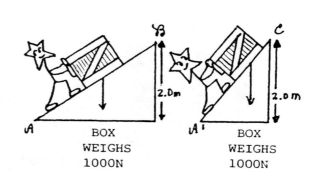

BOX WEIGHS 1000N   BOX WEIGHS 1000N

Since in both cases the output work is equal then the input work must be equal.

	INPUT WORK FROM A TO B	=	INPUT WORK FROM A' TO C

However,

	THE DISTANCE FROM A TO B	>	THE DISTANCE FROM A' TO C

This means

f X D   =   F X d
less X greater   =   more X shorter
force X distance   =   force X distance

The greater incline from A' to C results in a shorter distance which means more force must be applied. If the slope is not as steep such as from A to B, then less force is needed for the greater distance. This may explain why carrying packages up a steep hill is more difficult than on a slightly sloped hill.

POWER

Work says nothing about time. If a job can be done by one person in a given amount of time and can be done by a different person in less time, then we say the second person has more power. There is no difference in the amount of work done, if a job is done in 20 seconds or the exact same job is done in 1 minute. Time is the variable. A more powerful engine can do a given job more rapidly than a less powerful engine. The rate at which work is done is called power and is defined as follows:

$$POWER = \frac{WORK}{TIME}$$

Since work is measured in joules and time is measured in seconds, then power which is work divided by time is measured in joules per second and is referred to as watts. One watt of power is expended when 1 joule of work is done in 1 second. The unit for power is named after James Watt, the 18th century developer of the steam engine. Horsepower was the term originated by James Watt as the standard unit of power and was used by him to characterize the rate at which a horse did work. The horsepower as a unit of power was equivalent to moving a one pound object through a distance of one foot in one second. In the S.I. system, the unit of power is not the horsepower but is referred to as the watt.

In the U.S. we rate engines in units of horsepower (HP) and electricity in kilowatts (1 kilowatt = 1000 watts), but either may be used since they both refer to power. One horsepower is the same as 0.746 kilowatt or 1HP = 746W.

British System	S.I. System
Power = Work/Time	Power = Work/Time
= (Force x Dist)/Time	= (Force x Dist)/Time
= lb·ft/sec	= N·m/sec
= ft·lb/sec	= J/s
= HP	= W

In comparing the two systems, 1HP = 33,000ft·lb/min

$$1HP = 550ft·lb/sec$$
$$1HP = 746W$$
$$1HP = 746N·m/sec$$
$$550ft·lb/sec = 746W$$
$$1ft·lb/sec = 1.36W$$

When we pay our electricity bill, we are paying for kilowatt hours for the following reason:

Power = Work/Time

Multiplying both sides by Time to clear fractions gives us:

Power x Time = Work

or   Work = Power x Time

Since power is measured in kilowatts and time in hours, we have

Work = kilowatt x hrs

Thus, a kilowatt hour is a unit of work or energy.

Horsepower is still a common unit. If a 2HP engine is compared to a 1HP engine, the 2HP engine can do the same work in half the time or twice as much work in the same time.

ENERGY

When work is done energy is acquired. If 5J of work is done, then 5J of energy is acquired. This acquired energy could now be used to do other work.

There are many forms of energy. These energy forms include mechanical, heat, electrical, nuclear, light, sound, magnetism and many others. One of the major divisions of energy is mechanical energy within which there is potential energy and kinetic energy.

POTENTIAL ENERGY

Potential energy is stored energy or energy of position or condition. Any stored energy has potential for doing work. Water held back behind a wall has the potential to do work once it is released and thus it has potential energy. A stretched or compressed spring has potential energy. Chemical energy is a form of potential energy that can do work through a chemical reaction. Dynamite has potential energy because it can do work when it explodes. Potential energy is then the energy possessed by an object and is acquired because of work done to bring the object to its position or condition. Work is not energy itself but it is a transfer of energy. If work is done, energy is

energy is measured by joules.

acquired which in turn can be used to do other work.   Therefore, energy, like work, is measured in joules.

A unique form of potential energy is energy brought about because of an elevated position of an object and is called gravitational potential energy.   The energy is acquired because of the work done opposing gravity in bringing the object to its elevated position relative to its original position. Work is done in bringing the object to this elevated position and work is equal to force times the distance through which the force acts.   The upward force is equal to the weight of the object and the distance is equal to the height to which the object is raised.   This gives us the following formula:

fig. 5

GRAVITATIONAL POTENTIAL ENERGY = WEIGHT x HEIGHT
$$G.P.E. = mgh$$

because work = fd
weight = mg
and height = h
(fig. 5)

Suppose a 10N block falls freely from rest from a point 4.0m above the surface of the Earth (fig. 6).   With respect to the surface of the Earth, what is the gravitational potential energy of the block after it has fallen 1.0m? Ignore air resistance.

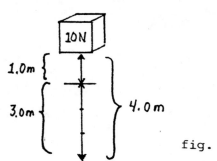

fig. 6

G.P.E. = Weight x Height
= (10N)(3.0m)
= 30J

(Note: The height is 3.0m because the block fell 1.0m of the original height of 4.0m.)

The gravitational potential energy gained or lost by a given object is always measured by the change in height between the initial height and the final height.   It does not matter what path the object took to change that height by either a gain or loss (fig. 7).

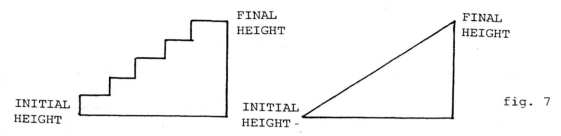

FINAL HEIGHT

INITIAL HEIGHT

FINAL HEIGHT

INITIAL HEIGHT -

fig. 7

In the formula G.P.E. = mgh the value of "g" is assumed to be constant and it is correct when used for calculations of objects involving heights near the Earth's surface. If calculations involve the lift off of rockets and their payloads to heights significantly above the Earth's surface, then the use of "g" is no longer accurate. The gravitational force between two objects depends on the distance between them as stated in the universal law of gravitation. As a rocket continues to move to greater heights, the distance changes and so does the gravitational force. This would produce a slightly different formula for G.P.E.

KINETIC ENERGY

If an object is in motion, it is capable of doing work and this energy of motion is called kinetic energy. An arrow moving through the air has kinetic energy. Heat involves kinetic energy of molecules. The kinetic energy of molecules moving in rhythmic patterns is necessary for sound. Electricity involves the kinetic energy of electrons. Wind, which is moving air, can do work by turning windmills while stationary air cannot.

The amount of kinetic energy an object possesses depends on both the mass of the object and the speed of the object. If two cars are moving with the same speed on the same road but with different masses, then the heavier car will have the greater amount of kinetic energy. Now suppose we have two cars with the same mass traveling on the same road but with different speeds. The car moving at the greater speed will have the greater amount of kinetic energy (fig. 8).

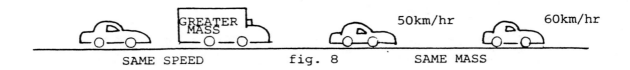

SAME SPEED   fig. 8   SAME MASS

In order for an object to acquire kinetic energy, work must be
done. Since work is equal to force times distance and since
kinetic energy depends on the work done then the following
formula can be derived for kinetic energy:

$$K.E. = \tfrac{1}{2}mv^2$$

The velocity squared has the effect that if the velocity is
doubled then the kinetic energy is quadrupled. In other words
twice the velocity equals four times the kinetic energy and the
object can do four times the work.

If the work done on an object goes "only" into changing the speed
then it could be said that the kinetic energy acquired equals the
work done on the object. Work may be done to a stationary or
moving object by either increasing or decreasing its speed. Once
a stationary object acquires speed it also acquires kinetic
energy and now, the moving object, can do work. Work is also
done in bringing a moving object to a halt.

If two cars with equal masses are traveling on the same road, one
with twice the velocity of the other, then the faster car will
have four times as much kinetic energy as the slower car. For
example, suppose the slower car is traveling at 40 km per hour
and the faster car is traveling at double that velocity or 80 km
per hour. Since the formula for kinetic energy is one-half the
product of the mass and the velocity squared and since the mass
is the same for both cars the only thing to be concerned about is
the velocity. Arithmetically, we could see that forty squared is
1600 and eighty squared is 6400 which is four times 1600.
Algebraically, if we let v = 40 and 2v = 80, then squaring v
gives us v^2 and squaring 2v gives us $4v^2$. In both cases we see
that the faster car has four times the velocity squared
and therefore, has four times the kinetic energy. Work stopping
the car requires a force and a distance to stop the car. In the
case of the faster and slower

cars with equal masses, if you press on the brake in the same way in each car, then the car with four times the kinetic energy will require four times the distance to stop. In both cases, the cars were brought to a halt. Thus, we see that the kinetic energy increases as the square of the velocity. If the velocity in the preceding example was three times as great then the kinetic energy would be nine times greater and the stopping distance for the same braking force is nine times as great.

KINETIC ENERGY VS MOMENTUM

momentum = mass x velocity

In Chapter VI which dealt with momentum, the concept of conservation of momentum was discussed and an illustration was given using the stand of five marbles as shown on the right (fig. 9). It was established that if two marbles are pulled out to the left and released then two marbles would

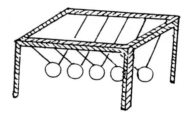

fig. 9

swing out on the right with the same initial velocity. If 2mv (left) equals 2mv (right), then the question arises as to whether 2mv (left) could equal m(2v) (right). That is, could we release two marbles from the left with a given velocity and have one marble swing out on the right with two times the velocity? The answer is no. Even though the momentum is the same before and after (2mv = m(2v)), the same is not true for the kinetic energy. The kinetic energy before would not be equal to the kinetic energy after. There would be an increase in kinetic energy which cannot happen as is discussed in the topic on conservation of energy later in this chapter.

K.E. IN FROM THE LEFT	K.E. OUT FROM THE RIGHT
$= \frac{1}{2}(2m)v^2$	$= \frac{1}{2}m(2v)^2$
$= mv^2$	$= \frac{1}{2}m(4v^2)$
↑These are not the same →	$= 2mv^2$

MOMENTUM IN FROM THE LEFT	MOMENTUM OUT FROM THE RIGHT
= 2mv	= m(2v)
= 2mv These are the same.	= 2mv

Note: If the marbles were allowed to swing back the momentum would not be conserved because the direction would change. The above example makes use of only one run from left to right.

Momentum is a vector quantity because velocity has direction and although mass is a scalar quantity, the product of a scalar and a vector quantity is still a vector quantity. However, with respect to kinetic energy the velocity, the vector quantity, is squared and becomes a scalar quantity. As a scalar quantity it cannot be cancelled.

Suppose a person is in the path of two running individuals and could only duck out of the path of one of them. One of the running individuals is twice as heavy as the other individual but is also running at half the speed. Notice they both have the same momentum.

	HEAVIER RUNNER	**LIGHTER RUNNER**
Momentum:	= (2m)v	= m(2v)
	= 2mv	= 2mv
	TWICE THE MASS	TWICE THE VELOCITY

However, the kinetic energy is not the same.

	HEAVIER RUNNER	**LIGHTER RUNNER**
K.E.:	= ½(2m)v^2	= ½m(2v)^2
	= mv^2	= ½m(4v^2)
		= 2mv^2

Even though they both have the same momentum the lighter but faster runner has twice the kinetic energy and can do twice the harm to you.

In   Chapter VI reference was made to elastic and inelastic collisions. Suppose a freight car moving at 20km/hr hits a similar freight car of equal mass which is stationary and therefore has a velocity equal to zero km/hr (fig. 10). When the collision takes place, there may be the sound of a crash and no damage to the trains as in an elastic collision. If it is an inelastic collision, there may also be

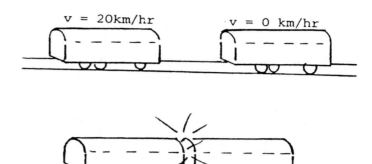

fig. 10

deformation of the trains. The total momentum remains the same, but the amount of kinetic energy decreases. This decrease in kinetic energy is not due to a loss of kinetic energy but rather to a transformation of one form of energy into another form. Energy is used to make the sound of the crash or to cause the deformation. The following discussion explains why the momentum remains the same while the total kinetic energy does not. Note that the total amount of energy remains the same but it is not all kinetic energy. Momentum is a vector quantity and the total momentum before the interaction must equal the total momentum after the interaction. The kinetic energy is a scalar quantity and cannot be cancelled. It, however, can be transformed into another form of energy other than kinetic energy. For example, if two trains moving toward each other collide and if they have equal and opposite momenta, then they would come to a stop resulting in zero momentum. This is because the total momentum of the trains before the interaction combines to zero and after it is still zero. The kinetic energy, however, of the two moving trains before the interaction cannot equal zero because kinetic energy is a scalar quantity which means it has magnitude without direction. Two scalar quantities cannot be cancelled when added. When the trains collide, their kinetic energy becomes zero because the velocity is zero for the stopped trains. Since the total energy before the interaction must be the same as the total energy after the interaction, the kinetic energy transforms into another form of energy such as sound or heat.

EXAMPLE:

Suppose a toy train rolling with a velocity of 10m/s on a straight track, locks with a stationary toy train and they roll together. How does the kinetic energy before the interaction compare to the kinetic energy after the interaction (fig. 11). Ignore friction.

v=10m/s   v=0m/s   v=5m/s

fig. 11

train 1   train 2   train 1 & train 2

$$momentum_{before} = momentum_{after}$$
$$mv_{before} = mv_{after}$$

momentum train 1 + momentum train 2 = momentum together

$$mass(10m/s) + mass(0m/s) = 2mass(v) \quad \text{(double mass)}$$
$$10m/s + 0m/s = 2v \quad \text{(divide mass)}$$
$$10m/s = 2v \quad \text{(combine like terms)}$$
$$5m/s = v \quad \text{(divide by 2)}$$

Now they are both moving with ½ the velocity of the original train

$$K.E._{before} = K.E._{train 1} + K.E._{train 2}$$
$$= \tfrac{1}{2}mv^2{}_{train 1} + \tfrac{1}{2}mv^2{}_{train 2}$$
$$= \tfrac{1}{2}mass(10m/s)^2 + \tfrac{1}{2}mass(0m/s)^2$$
$$= 50J \text{ *}$$

$$K.E._{after} = K.E._{(train 1 + train 2)}$$
$$= \tfrac{1}{2}mv^2{}_{(train 1 + train 2)}$$
$$= \tfrac{1}{2}(2mass)(v)^2$$
$$= mass(5m/s)^2$$
$$= 25J \text{ *}$$

* The quantity for "mass" has been left off because it is the same in the final solution of K.E. before & after and only the 50J & 25J is needed for a comparison.

Now the kinetic energy is ½ the original amount of which some was converted into heat, some into sound and some into deformation, etc.

CONSERVATION OF ENERGY

In the marble example discussed earlier, if two marbles were released from the left with a given velocity, one marble could not leave the right side with twice the velocity because the kinetic energy would be greater. It would be impossible because energy cannot be created nor can energy be destroyed. Energy may be transformed from one form into another, but the total amount of energy never changes. This concept is the foundation of the law of conservation of energy. If there is a conversion of energy in the system from one form to another, the total energy before the conversion equals the total energy after the conversion. Energy is never lost. It may be converted from one form into another form where the second form is useless. For example, some of the energy used to make toast browns the bread and some of that energy just heats up the room and becomes useless energy. Some of the kinetic energy of a falling rock becomes the sound you hear when the rock strikes a surface and some of the energy is changed to heat. Where there is friction, the amount of energy used to overcome the friction is equal to some of the loss of kinetic and potential energy from the system.

TOTAL ENERGY BEFORE = TOTAL ENERGY AFTER

Suppose we consider an ideal system where energy is not lost to friction or air resistance.
If a pendulum bob is released
from rest, it will rise no
higher than its starting point
when it swings to the other
side. At the starting point
the pendulum bob, has potenetial
energy. As the bob swings to
the other side it loses potential
energy which is converted into
energy of motion or kinetic energy.
At the lowest point the bob has
lost its potential energy and has
its maximum kinetic energy. The
kinetic energy gained equals
the potential energy lost. When

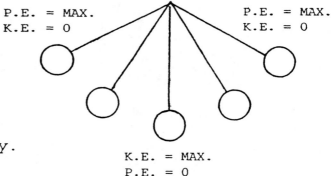

P.E. = MAX.
K.E. = 0

P.E. = MAX.
K.E. = 0

K.E. = MAX.
P.E. = 0

fig. 12

the bob goes to the other side it continues to swing until it reaches its original height, at which point the kinetic energy has completely converted back to potenetial energy (fig. 12). If friction is not ignored, the bob could not reach its original height, even though the total amount of energy does not change. Only the form of the energy changes and some of the energy is

used to overcome friction. In some systems, work disappears
entirely into friction and no kinetic energy or potential
energy remains. When a person works with sandpaper the work
input is converted to other forms of energy of which the most
obvious form is heat.

Nuclear energy or fossil fuel energy is transformed into heat
which is then used to produce steam that turns a turbine. The
mechanical energy of a turbine is used to power up a generator
which gives us electrial energy. Electrical energy can be
transformed into heat or light or even mechanical energy that can
do other work. Engines of all kinds convert thermal energy of
fuel into kinetic energy that drives machinery. The nice thing
is that there is a formula to calculate each type of energy.

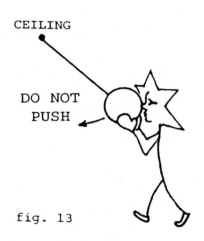

CEILING

DO NOT
PUSH

fig. 13

If a ball is attached to a string
and suspended from the ceiling,
then it could be pulled to the
right, allowed to swing to the left,
and back again to the right like
the pendulum in the previous
illustration. If the ball were
released from just below your nose,
it would never hit you in the face
on the return swing. It is
important to make sure you just
release the ball and do not push
the ball (fig. 13).

EXAMPLES FOR THOUGHT IX

1. The rate of change of work with respect to time is called
    (1) energy   (2) momentum   (3) force   (4) power

2. The work done in holding a weight of 50N at a height of
    5m above the floor for 5 seconds is
    (1) 0 J   (2) 50 J   (3) 250 J   (4) 1250 J

3. If the velocity of a moving object is doubled, the object's kinetic energy is
   (1) unchanged   (2) halved   (3) doubled   (4) quadruled ✓

4. If a 10kg mass is raised vertically 2 meters from the surface of the Earth find its gain in potential energy.   G·m·h   196
   $10 \times 9.8 \times 2 = 196 J$

5. What is the kinetic energy of a 5kg ball traveling at 10.5m/s?

   $\frac{1}{2} \cdot 5 \cdot 10.5^2$ 275.625 J
   $K.E = \frac{1}{2} m v^2$

6. In the diagram at the right a pendulum bob is released from point A and swings freely through point B. Ignoring air resistance and friction compare the bob's kinetic energy at point A to its potential energy at point B. It is
   (1) half as great   ✓(3) the same
   (2) twice as great   (4) half as much

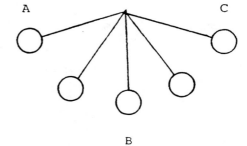

7. A satellite travels around the Earth in an elliptical orbit going from point A to point B as shown in the diagram at the right. Using the information from Kepler's second law which says that as a planet travels around the Sun it sweeps out equal areas in equal time intervals, determine the changes that occur in the potential energy and kinetic energy of the satellite as it moves from point A to point B.

   (1) both potential energy and kinetic energy increase
   (2) both potential energy and kinetic energy decrease
   ✓(3) potential energy decreases and kinetic energy increases
   (4) potential energy increases and kinetic energy decreases

8. A 10kg mass and a 20kg mass are released simultaneously from a height of 100 meters above the ground. If they fall for 5 seconds, which of the following will be different

for the two masses?
(1) acceleration (2) speeds (3) k.e. (4) displacement

9. If a free falling object is dropped from an elevated point,
    when will the kinetic energy of the object equal the
    potential energy of the object.
    (1) at the start of the fall
    √(2) halfway between the start and the end of the fall
    (3) at the end of the fall
    (4) at all points during the fall

10. How does the stopping distance of a car going 90mph compare
    to the stopping distance of the same car going 45mph using
    the same braking force?

MATH HUMOR IX

1. How far do you have to count before using the letter **"A"**
    in the spelling of a number?

2. Pick any three digit number.
    Reverse the digits.
    Subtract the smaller number from the larger number.
    If you tell me the right most digit, I'll tell you your
    answer.

3. Pick any three digit number.
    Reverse the digits.
    Subtract the smaller number from the larger number.
    Reverse the digits in your answer.
    Add this new number to your last answer.
    Your answer is...

4. Arrange the numbers 1 through 9 so that they add up to
    99,999.

5. Without using a pencil or paper find the number of nines
    in the numbers from 1 to 100.

6. Use the numbers from 0 to 9 inclusive to fill in the blanks
    and to equal 100.

    __ X __ + __ + __ + __ + __ + __ + __ + __ + __ = 100

Chapter X

HEAT AND TEMPERATURE

One cold winter day my son came in from playing in the snow complaining his fingers were cold. I told him to blow on his fingers and they'll warm up. While he was blowing on his fingers, I prepared some hot soup for him. When I gave him the soup he said it was too hot. "Blow on the soup and it'll cool off", I told him. Now he wondered how blowing could warm something up and also cool something off. This leads to the following discussion of how heat travels and why.

Heat travels from one place to another. It travels from a region of higher temperature to a region of lower temperature. If a hot object and a cold object are placed together, heat will flow from the hot object to the cold object until they both have the same temperature.

Heat which is thermal energy travels in one of three ways.

> 1. conduction
> 2. convection
> 3. radiation

In each case the amount of thermal energy is the same before and after it moves. When thermal energy is transformed into mechanical energy, it then has the ability to do work. The study of this transformation is called THERMODYNAMICS.

CONDUCTION

Try standing dominoes on their side about one half inch apart. Tap the last domino and see what happens to all the dominoes (fig. 1). The way in which all the dominoes fall, one after

fig. 1

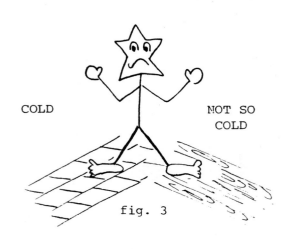

fig. 2

the other, is similar to the way in which heat travels by the method known as conduction. Place a spoon into a coffee cup and then pour in hot coffee (fig. 2). Notice that slowly the handle end of the spoon begins to get hot. This is because the molecules of the portion of the spoon in the coffee begin to move more rapidly. These molecules collide with their neighbors causing them to move faster and so on. The increased molecular motion is carried from molecule to molecule throughout the entire spoon.

How well an object conducts heat depends on the bonding of the molecules. Solids are very good conductors of heat and metals are the best solid conductors. Of the metals, silver is the best conductor, copper is next, and then there is aluminum and iron. If a metal object and a wood object are both standing in the same room they must both be at the same temperature. Yet, the metal object would seem to be colder to an individual touching both. They are both at the same temperature, but the metal is a better conductor of heat and conducts the heat away from the individual's fingers faster than the wood. The temperature in the fingers is higher than that of the object but because heat travels from a region of higher temperature to a region of lower temperature the heat will leave the fingers making them feel cold. The metal object causes the heat to leave the fingers much more rapidly than the wood object. A similar situation exists when stepping on a tile floor versus stepping on a wood

COLD

NOT SO COLD

fig. 3

floor in the same room (fig. 3). Tile is a better conductor of heat and, therefore, conducts the heat away from the foot more rapidly.

Metals, such as aluminum and copper, are used as cooking utensils because they are good conductors of heat. The handles are made of wood or plastic because they are poor conductors. Try lifting a hot pot with a metal handle as opposed to a wood or plastic handle. You might get burned if you didn't use oven mittens. Wool, plastic, straw, paper cork, and styrofoam are other examples of poor conductors of heat (fig. 4).

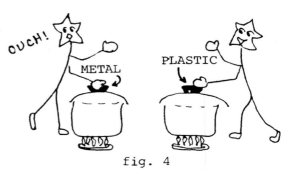

fig. 4

Liquids and gases are also poor conductors. Air is a poor conductor of heat which means that a porous substance with a large number of small air spaces is a poor conductor of heat. But, if it is a poor conductor of heat, it is also a good insulator. The air spaces in wool, fur, or thermal garments, help to prevent loss of body heat and in so doing keeps an individual warm on a chilly day. Layers of clothing serve the same purpose in the same way. Snow is a poor conductor and does the same thing for the Earth, in that it keeps it warm by preventing heat from escaping. Remember, heat travels from a region of higher temperature to a region of lower temperature. Insulation is used in order to prevent the heat from escaping.

Molecules of all substances are moving all the time. Molecules of a liquid are freer to move than those of a solid and molecules of gases are even freer to move than those of a liquid. Heat is the kinetic energy or the energy of the motion of these molecules.

CONVECTION

Heat travels in liquids and gases by a method which uses currents and this method is called convection. When air is heated it immediately rises and is replaced by cool air. After the warm air cools it moves downward and the whole cycle begins again. There are several explanations as to why hot air rises. One reason has to do with the fact that gravity causes the molecules

-114-

closest to the Earth's surface to be more dense. Suppose each molecule close to the surface of the Earth has the same kinetic energy. Now if one molecule is heated which means it acquires more kinetic energy or energy of motion, it will move upward in the direction of fewer molecules where the air is less dense and the molecule has room to move. The molecule gives up some of its kinetic energy when it hits another molecule and sets the other molecule into motion. As the first molecule continues to give up its kinetic energy in colliding with other molecules, it has less energy of motion. It is pulled back closer to the Earth's surface, ready to be heated up again. The molecules collide more often because they are heated and have energy of motion as opposed to getting heated because they collide.

The rising of hot air is most obvious at the seashore. In the daytime the sand is hotter than the water and consequently the air over the sand gets heated and rises. The cooler air from over the water rushes in to take its place producing a sea breeze (fig. 5). At night the sand cools off more rapidly and since the water is now warmer the air above the water rises causing the air above the land to replace it resulting in a land breeze (fig. 6).

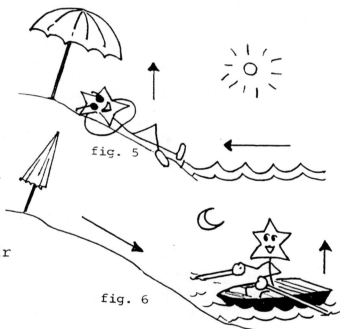

fig. 5

fig. 6

RADIATION

The third method by which heat travels is radiation. Radiant energy travels as electromagnetic waves and does not need a medium through which to travel. It is absorbed by dark rough objects, reflected by shiny smooth objects and transmitted by transparent objects. The heat from the Sun is transformed into radiant energy which travels through the atmosphere. After the energy passes through the atmosphere, it may be absorbed by objects at the Earth's surface resulting in a warming of these objects. In other words, only energy, not heat, travels from the

Sun and this energy is transformed back into heat when it strikes an object or surface. The energy travels through space in a method described as a wave and at the point of contact with an object or surface this tranported energy is converted into internal energy of the absorbing material. Radiation is the transmission of radiant energy and not heat. Further detailed discussion on the nature of the radiant wave energy and electromagnetic waves will follow in a later chapter.

RADIOMETER

A radiometer is an instrument which is designed to measure radiant wave energy. A radiometer consists of a glass bulb which contains four vanes that rotate about a center shaft (fig. 7). Almost all of the air is removed. Radiation is the key to the movement in the radiometer. The vanes in the radiometer are alternately dark and light in color. When the energy from the Sun or other light source strikes the vanes, the dark vanes absorb the radiant wave better than the light vanes which reflect it. The radiant wave energy is transformed into heat when it strikes the dark vane's surface and the dark vanes become warmer. When the molecules of air inside the radiometer come into contact with these warm vanes, they absorb energy and rebound with higher speeds than those meeting the light vanes. According to Newton's Third Law there is an equal and opposite reaction. This would result in the dark vanes moving in the direction as though they were pushed in that direction. As the light source gets stronger and stronger, the vanes revolve about the center more and more rapidly. The speed with which the vanes of the radiometer spin shows the intensity of the incoming radiant energy.

fig. 7

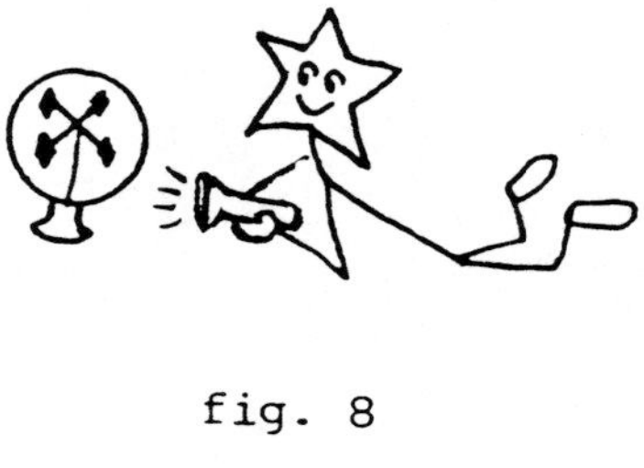

fig. 8

TRY THIS EXPERIMENT:

Place a radiometer on the windowsill in
your house or office. Watch what happens
on days when the Sun is strong as opposed
to cloudy days when the Sun is not shining.
Or try focusing a flashlight at the radiometer
and watch the vanes spin (fig 8).

THERMODYNAMICS

Thermodynamics is the study of thermal energy which is transformed
into mechanical energy and as such does work in steam engines,
gas engines and other types of engines. Energy is always being
transformed from one form into another form but it is never used
up. The total amount never changes. If it is never used up,
then why do we have to keep drilling for oil or looking for
other sources of energy. The answer is that not all forms of
energy are equally useful. Thermal energy is only useful at very
high temperatures. Very hot steam is needed to power up a
generator. This concept is summarized in the FIRST LAW OF
THERMODYNAMICS which says that the total amount of energy does
not change. If a certain amount of energy is added to a system,
the total amount of all output energy from that system is the
same as the amount initially put into the system. More energy
cannot be produced and energy cannot be used up . Basically,
this statement is applying the law of conservation of energy to
thermal energy. However, as energy is converted from one form
into another form, it is not done with 100% efficiency. This
process of converting energy results in an output of a large
percentage of low temperature thermal energy which cannot do
work. This is the SECOND LAW OF THERMODYNAMICS which says
nothing is 100% efficient. No machine is completely efficient in
converting energy into work. Thus, there is a need to keep
looking for energy sources. The following example shows how
energy may be converted from one form into another form and how
some of the energy gets wasted.

    Example: 1. Hot steam turns a turbine which runs an electric
                generator. Some energy leaves in the form of
                useless warm water.
             2. The electric generator loses energy due to
                friction.
             3. The electricity from the generator may be used
                to make toast with some of the energy just warming
                the environment.

At each step some energy becomes useless.

TEMPERATURE VS HEAT

TEMPERATURE:

The measure of the average kinetic energy of the molecules of a substance or material.

HEAT:

The total amount of kinetic energy of the molecules of a substance or material.

Heat and temperature are related. Heat or thermal energy refers to the kinetic energy of the molecules. As the internal energy increases, the molecules move more rapidly and the temperature which is a measure of this kinetic energy rises. If the material loses heat, in other words, if the molecules slow down causing the kinetic energy to decrease, then the temperature goes down.

Temperature depends upon the average kinetic energy of the molecules. Since a large amount of water has more molecules than a smaller amount of water, the larger amount of water will require a greater amount of heat input to bring the many more molecules to the same energy level as the smaller amount of water to achieve the same temperature. If the average kinetic energy of the larger amount of water was less then its' temperature would be less. When a cup of coffee is hot, the molecules in it have a higher average kinetic energy than when the coffee is cold. A two quart pitcher of hot coffee has twice the kinetic energy as a one quart pitcher of coffee at the same temperature. There is twice as much heat energy in the two quart pitcher of coffee as is in the one quart pitcher, but the temperatures are the same.

COLD COFFEE. NOT ENOUGH K.E.

THE THERMOMETER

A thermometer is used to measure temperature or average kinetic energy. The most common thermometer is made of glass with an enclosed liquid such as mercury which expands more rapidly than the glass. Three commonly known scales such as the Celsius, the Fahrenheit and the Kelvin, are used to indicate the temperature. Once a point is determined for the freezing point and another point is chosen for the boiling point then equal intervals are marked between those points. Both the Celsius and the Fahrenheit were designed in this manner. The Celsius scale was named for Anders Celsius, a Swedish astronomer, who introduced this scale formerly known as the centigrade scale. The Fahrenheit scale was named for Gabriel Fahrenheit, a German physicist, who made a mercury thermometer and introduced this scale. The Kelvin scale was introduced by Baron Kelvin, a British physicist and mathematician.

In order to calibrate the Celsius scale the thermometer was put into an ice water mixture and when the mercury stopped falling, this thermal equilibrium point was marked as 0°C. Then the thermometer was put in boiling water at a pressure of one atmosphere and when the mercury reached thermal equilibrium, the scale was marked as 100°C. On the Fahrenheit scale the freezing point was marked as 32°F which is an arbitray starting point and 212°F is the boiling point. Once the freezing and boiling points were selected the amount between them was divided into equal intervals. The intervals on the Celsius and the Fahrenheit scales are not the same.

$$1°F = (5/9)°C$$

This means that every time the Fahrenheit scale jumps up 1°F, the same temperature increase will be a (5/9)°C jump on the Celsius scale.

Kelvin believed that when no more energy can be extracted and the temperature cannot go lower, then this limiting temperature should be zero or better known as absolute zero. At this point all of the molecules have lost their kinetic energy. In this way zero means no more movement. Absolute zero is the lowest possible temperature in the Universe and it corresponds to approximately (-273°C) on the Celsius scale and to approximately (-459°F) on the Fahrenheit scale. The intervals on the Kelvin and Celsius scale are the same.

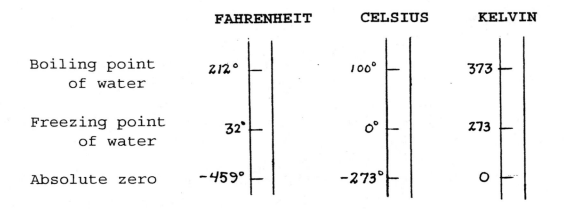

	FAHRENHEIT	CELSIUS	KELVIN
Boiling point of water	212°	100°	373
Freezing point of water	32°	0°	273
Absolute zero	-459°	-273°	0

There exist formulas to convert from one scale to another.

$$F = \frac{9C}{5} + 32$$

For converting
Celsius to
Fahrenheit

$$C = \frac{5(F - 32)}{9}$$

For converting
Fahrenheit
to Celsius

FAHRENHEIT

-112°   -40°   14°   32°   77°   98.6°   212°   248°

---+-------+-----+----+-----+-----+-----------+-------+--------

-80°   -40° -10°   0°   25°   37°   100°   120°

CELSIUS

If the two formulas are   difficult to use when traveling and it
is important know the Fahrenheit equivilent when Celsius is given
and vica versa, here is an easy method to get a ball park
figure. When changing Fahrenheit temperature to Celsius
temperature just subtract 30 and then take half.   When changing
Celsius temperature to Fahrenheit temperature double the number
and add 30.

EASY METHOD:
GIVEN FAHRENHEIT: subt. 30 and divide by 2
GIVEN CELSIUS: mult. by 2 and add 30

TRY THESE EXAMPLES:

Example 1:   Change 68°F to Celsius.

EASY METHOD:

Subtract 30   68-30=38

Divide by 2   38/2=19°C

ACTUAL FORMULA:

The actual answer is C = $\frac{5}{9}$(68 - 32)

$$= \frac{5}{9}(36)$$

$$= 20°C$$

Example 2:   Change 15°C to Fahrenheit.

EASY METHOD:

Multiply by 2   15 X 2 = 30

Add 30   30 + 30 = 60°F

ACTUAL METHOD:

The actual answer is F = $\frac{9(15)}{5}$ + 32

$$= 27 + 32$$

$$= 59°F$$

The easy method is close enough if you are traveling in a country where you are not familiar with the temperature readings and you want to know whether you should or should not take a jacket.

burr!!!

EXPANSION

As heat is added to any material, the internal kinetic energy increases. The molecules move faster and faster and move farther apart. Consequently, the material expands. Although liquids and solids expand, they expand at different rates. Suppose two metals which have different rates of expansion are welded together to form a straight strip of metal at room temperature. If heat is added the metal strip called a bimetallic strip, will bend in the form of an arc with the metal that has the greater rate of expansion on the outside (fig. 9). In older thermostats, a heated bimetallic strip arcs away from a contact

fig. 9

thereby preventing the circuit from being completed. The boiler is turned off automatically until the bimetallic strip cools and straightens out. The bimetallic strip touches the contact, completes the circuit, and causes the boiler to go on again.

Did you ever have a tight metal lid on a glass jar that would not come off? Try holding the top of the jar under hot water. The lid will loosen because metal expands at a faster rate than glass.

TRY THIS EXPERIMENT:

Here is a classic experiment in physics. Suppose you have a sheet of metal and you cut out a hole in the center. If you heat the metal, will the hole get smaller or will it get larger? The answer is that the hole will get larger because metal expands when heated. Try drawing a circle on the metal instead of cutting out the circle. You can see that the drawn circle gets larger when the metal is heated, not smaller. A similar example consists of taking eight metal squares and arranging them as shown in fig. 10 at the right. The middle does not have a metal square. Take the metal squares and heat them. Now set them out in the same way as they were before (fig. 11). What happens to the square space in the middle? Does it get smaller or larger? This experiment shows that the center space gets larger not smaller.

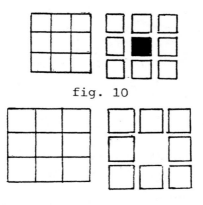

fig. 10

fig. 11

CONTRACTING

It is widely believed that as things cool they contract. Water contracts while cooling until a certain point and then it expands. The critical point at which water stops contracting and begins to expand is at 4°C. This does not mean that water will freeze at this point. The water will begin to expand at 4°C and continue to expand as the temperature drops until it freezes at 0°C. This explains why the ice cube tray should not be filled to the top. Also, the expansion means that the ice cubes in a pitcher of water are less dense than the water and therefore they float. The expansion also explains why only the top layer of water in a lake freezes in winter. As the water cools from 4°C the colder less dense layers come to the top where the top layer, at 0°C, freezes.

RATE OF COOLING

Newton found that the rate of cooling was not constant and came up with his theory on the rate of cooling. He found that if the difference in temperatures is very great then the rate of cooling is also great but if the difference in temperature is not so great then the rate of cooling is less. The rate of cooling directly relates to the difference in temperature. Remember, heat travels from a region of higher temperature to a region of lower temperature. If an object is hotter than its surrounding, heat will travel from the object until the object's temperature is in equilibrium with the surrounding temperature. All along, the difference in temperature is getting smaller and smaller and, according to Newton, the rate of cooling will also slow down. If you add cream to your coffee when you get it, the difference between your coffee's temperature and the surrounding temperature will now be less and your coffee will cool at a slower rate than if you wait for your coffee to cool before putting in the cream.

EXAMPLES FOR THOUGHT X

1. Given the formula for Fahrenheit ($F = \frac{9C}{5} + 32$), derive the formula for Celsius.

2. Why does the quick method for finding the ball park figure for Fahrenheit and Celsius decribed in this chapter give an answer which is relatively close?

3. If the Celsius temperature is 37°C, find the corresponding Fahrenheit temperature.

4. Find the Celsius temperature when the Fahrenheit temperature is 86°F.

5. The internal energy of a substance is at a minumum when its temperature is
   (1) 0°C   (2) 0 K   (3) 273°C   (4) 459°F

6. When water cools from 6°C to 3°C it
   (1) contracts only (3) first contracts, then expands
   (2) expands only   (4) first expands, then contracts

7. The figure to the right represents a sheet of metal with a box cut out of the center. If heated will the center square region get smaller or larger?

8. In this chapter, two easy methods were given for approximating the change between the Celsius and the Fahrenheit scales. For what temperature does either approximation rule give the same value as the actual formula?

9. Find the Fahrenheit temperature when the Celsius temperature is -40°C.

MATH HUMOR X

1. Can you give an example of something that expands when heated and an example of something that contracts when turning cold?

2. Which travels faster: hot or cold?

Chapter XI

Suppose there is an alarm clock sitting on a table with a dome shaped glass covering it (fig. 1). It is set to ring at 3:00 p.m. and it does ring at that time. The ring could be easily heard. Do you think that, if the air were somehow pumped out, the ring would still be heard? The answer is no because audible sound waves need a medium through which to travel. They cannot travel through a vacuum which the absence of air would create. Audible sound waves travel better through some mediums than through others. For example, if you put your ear close to a railroad track and listen for an oncoming train, you'll be able to hear it sooner than listening for the sound through the air.

fig. 1

TRY THIS EXPERIMENT:

A HOMEMADE TELEPHONE

Use two small cans and a long string. Make a hole in the base of each can and pull each end of the string through, knotting it on the inside of each can. One person holds one can to his ear and the other holds the other can to his mouth (fig. 2). With the string pulled tight they can carry on a conversation by alternating the ear and mouth positions.

fig. 2

LONGITUDINAL VS TRANSVERSE WAVES

The audible sound waves that you hear are known as longitudinal waves. These longitudinal waves consist of vibrating particles of a medium in the form of alternating compressions and expansions (fig. 3). The vibration is parallel to the direction of the motion of the longitudinal waves. On the other hand, there is another type of waves such as light waves. These waves do not need a medium through which to travel and are called transverse waves (fig. 4). With respect to transverse waves, the particle vibration is perpendicular to the direction of the wave motion.

LONGITUDINAL WAVE fig. 3

TRANSVERSE WAVE fig. 4

For example, suppose you were floating in the ocean beyond the point where the waves break. A wave comes along and rolls to the shore but you just move up and down with the wave. Your movement is perpendicular to the direction of the wave motion. The wave is actually delivering energy to the shore but not the water. Another example would be to drop a pebble into a pond thus creating an outward ripple effect in all directions. The vibration caused by the pebble is perpendicular to the ripples. In a transverse wave, the particles vibrate perpendicular to the direction in which the wave travels as opposed to a longitudinal wave where the particles vibrate parallel to the direction in which the wave travels.

CHARACTERISTICS OF A TRANSVERSE WAVE

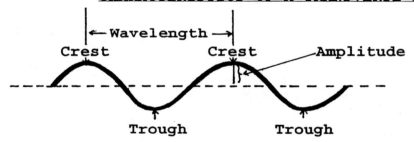

fig. 5

AMPLITUDE:

The distance a wave rises or falls with respect to a given base line. The amplitude depends upon the amount of energy used to create the wave (fig. 5).

WAVELENGTH:

The distance between two crests of a wave (fig.5).

FREQUENCY:

The number of waves that pass a given point in one second.

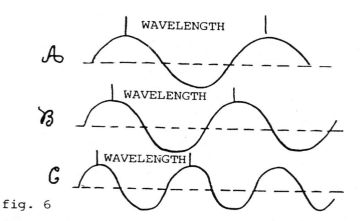

fig. 6

In fig. 6, wave B has a wavelength that is shorter than wave A. The wavelength of wave C is even shorter than that of either wave A or wave B. The frequency of wave C is greater than that of wave A or wave B. Wave C has more waves in a given time interval than either of the other two waves. In general, long waves have low frequencies and short waves have high frequencies (fig. 6).

All transverse waves travel at the same speed in a vacuum. That speed is 3×10^8 m/s. At that rate it takes approximately 8 minutes for the Sun's rays to reach us from almost 93 million miles away. There are many waves all around us, but most of them are invisible. Radio waves, infrared rays, and X-rays are some examples of invisible waves. The words "rays" and "waves" could be used interchangeably. Instead of saying infrared rays they could be called infrared waves. The various transverse waves have different wavelengths and different frequencies. If all the transverse waves are arranged according to their wavelengths and frequencies, they form what is called the electromagnetic spectrum (fig. 7). The word "electromagnetic" refers to the fact that these waves can be produced with the use of electricity or magnets.

Remember: 10^{10} = 1 followed by 10 zeroes
= 10,000,000,000
10^{-16} = a decimal followed by
15 zeroes and a one
= .0000000000000001

Gamma rays have such short wavelengths that hundreds of millions of their waves could fit across your thumbnail. The visible portion of the spectrum is very small. When the wavelengths of all colors appear together they are seen as white.

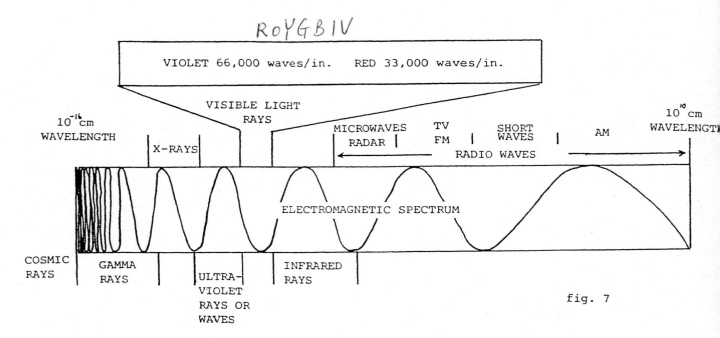

ROYGBIV

| VIOLET 66,000 waves/in.   RED 33,000 waves/in. |

10^{-16} cm WAVELENGTH

VISIBLE LIGHT RAYS

X-RAYS

MICROWAVES TV SHORT AM
RADAR | FM WAVES

10^{10} cm WAVELENGTH

RADIO WAVES

ELECTROMAGNETIC SPECTRUM

COSMIC RAYS   GAMMA RAYS   ULTRA-VIOLET RAYS OR WAVES   INFRARED RAYS

fig. 7

GAMMA
RAYS

-very high
 energy
 radiation
-used in
 medicine

ULTRA- VIOLET
RAYS OR WAVES
    -emitted by
     the sun and
     some man made
     lamps
    -wavelength
     range is
     above 66,000
     to more than
     2,500,000
     waves/in.
    -produces
     vitamin D
     in your skin
    -some
     substances undergo
     fluorescence when
     irradiated with
     UV light
    -kills bacteria
     and is used to
     purify foods and
     water
    -used by police
     and scientists
     for detecting
     blood stains,
     forged documents
     and faked oil
     paints
    -invisible ink:
     can be seen
     under UV lamp,
     otherwise it is
     invisible

INFRARED
RAYS

-you cannot see
 infrared
 radiation but
 you can feel it
-a toaster browns
 bread by
 infrared
 radiation
-infrared lamps
 are used to
 dry paints in
 industry
-used by military
 for detecting
 targets
-wireless remote
 controls for
 t.v.
-LASERS: some of
 the most powerful
 ones in use emit
 infrared light

MICROWAVES

-very short
 radio waves
-passes through
 rain, smoke
 and fog and
 is therefore
 used in
 communication
-used to send
 telephone
 messages in
 mountain areas
 where it is too
 difficult to
 put up
 telephone
 lines
-cooking
-radar

-128-

Just for your information: The word ROYGBIV, pronounced ROY-G-BIV, is an acronym used by art students to remember the order of the rainbow of colors. R = Red
                                                    O = Orange
                                                    Y = Yellow
                                                    G = Green
                                                    B = Blue
                                                    I = Indigo
                                                    V = Violet

When light falls on a black object, no colors are reflected and that is why the object appears as black. All of the colors are absorbed and turned into heat. It is one of the reasons why you may feel hot wearing black clothes in the summer. When light falls on white clothes, all of the colors of light are reflected which is why light or white clothing is worn in the summer Sun. The color, white, consists of the waves of all the colors taken together.

TOP SPINS

TRY THIS EXPERIMENT:
MAKING A SPINNING TOP:

Try building a top with a large nail and a circular disc. Divide the disc into seven sections as shown and color them according to the rainbow colors. Spin the disc and notice that when the waves representing all the colors get mixed together, they appear as white (fig. 8).

SURFACE APPEARS TO BE ALMOST WHITE

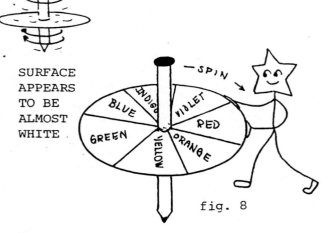

fig. 8

When transverse waves pass through glass or water, the speed is reduced. As white light travels through a prism the different wavelengths travel with different speeds causing them to bend. This separates the wavelengths producing a rainbow of colors (fig. 9). Water droplets in the sky, after a rain, have the same effect of separating the different wavelengths producing the rainbow in the sky. For the same reason, water droplets in the air at Niagara Falls combined with the sunlight help

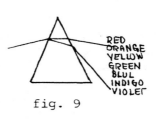

fig. 9

RED
ORANGE
YELLOW
GREEN
BLUE
INDIGO
VIOLET

to produce a constant rainbow at the falls. This leads us to the question, "Why is the sky blue?" The Sun's rays contain all of the wavelengths for all of the colors which together produce the color white. The sky appears blue because as the Sun's rays pass through the atmosphere the wavelength for blue scatters more than the other wavelengths.

THE GREENHOUSE EFFECT

Glass and certain other materials allow visible light to pass through, but absorb or reflect the longer wavelengths. A glass building will allow visible light to enter and to warm the interior. Incoming waves give up some of their energy to the interior objects. They now become waves with less energy and since the frequency is directly proportional to the energy, the frequency of each wave is now less. A lower frequency means a longer wavelength. These longer wavelengths fall into the infrared portion of the spectrum and cannot escape through the glass. Since waves enter and none leave, the warmth builds up which is one of the causes of the greenhouse effect. The interior of a car sitting in the Sun with the windows closed is warmed up in the same way. The accumulation of heat energy is built up when the shorter wavelengths pass through the glass and are converted into longer wavelengths when some heat energy is absorbed by the interior of the car. Again the longer wavelengths cannot radiate out through the windows and this causes the temperature in the car to rise.

PHOTON

How are transverse waves created? Basically, the answer depends upon knowing that the atom has electrons orbiting around a center called the nucleus. An analogy might be the Sun as the center and the planets as the electrons orbiting around the Sun. Another analogy could be satellites in orbit around the Earth. Using the second analogy, suppose you want to lift a satellite already in orbit to an orbit which is further from the Earth's surface. A significant amount of energy is required and this energy is the potential energy of the satellite. In the same way, an electron orbiting a nucleus absorbs energy when it moves to a higher energy level or

ORBITING
ELECTRON

NUCLEUS

an orbit further from the nucleus. When the electron falls back to its starting level, it gives off the acquired energy. This bundle of energy which is emitted when the electron returns to its original orbit is called a PHOTON of energy. The process of boosting the electron is called EXCITATION and the falling back of the electron is called DE-EXCITATION. The frequency of the wave emitted depends upon the size of the transition and this depends upon the energy causing the displacement. The electron's higher level is only momentary and it immediately returns to its original level at which point its' acquired energy is released as radiation or radiant wave energy. The frequency of the wave is related to the energy transition from the higher level to the original level. A photon in a beam of red light carries a certain energy that corresponds to the frequency for red.

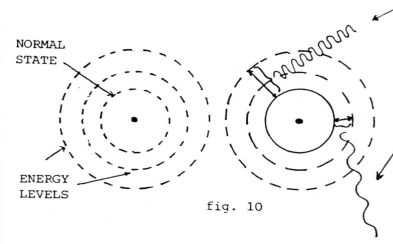

NORMAL
STATE

ENERGY
LEVELS

fig. 10

SHORTER WAVELENGTH

(greater transition
from the higher
level back to the
original level)
(fig.10)

LONGER WAVELENGTH

(smaller transition
from the higher
level back to the
original level)
(fig.10)

Radiant wave energy is emitted when electrons in an atom make a transition from a higher to a lower energy level and the frequency is directly proportional to the energy. E ∾ F, where the sideways "S" means "is proportional to".

In some situations, it is advantageous to describe the manner in which radiant energy travels as wave behavior and in some cases a description using particle behavior might be better. The speed with which the energy travels is 3×10^8 m/s in a vacuum.

NEON SIGNS

The light emitted by advertising
signs is a good example of
excitation and de-excitation process
for different gases which produce
different colors in the sign. The
sign is made of glass tubing shaped to
convey the required name or information. After the
electrodes are placed in the tubes, the tube is
filled with the gas needed to produce the required
color. If red is desired, then NEON gas is used. Almost all
such signs are called neon signs but they do not all contain neon
gas, if they do not emit the color red. When the sign is on the
electrons.in the neon gas are boosted to a higher energy level
and upon de-excitation, emit the wavelength corresponding to the
color red. Another gas would emit another wavelength and
consequently a different color. This process continues to occur
again and again.

INCANDESCENT LIGHTS

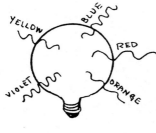

fig. 11

In ordinary incandescent lights, light is emitted
by gases such as mercury vapor. The tungsten
filament excites the electrons by heating them
with an electric current. The atoms containing
these electrons collide with one another and
many of them absorb additional energy causing
the orbiting electrons to jump to higher energy
levels. The orbiting electrons gain varying
amounts of energy and as they return to their original level they
release this acquired energy as waves or bundles of energy. The
frequency of the wave corresponds to the amount of acquired
energy. Light from an ordinary incandescent bulb contains all
the wavelengths for all the colors of the visible portion of the
spectrum and therefore, appears as white light. Since each color
radiates energy at a different frequency, the bulb contains a
jumble of frequencies (fig.11).

FLUORESCENCE

Some atoms are excited by absorbing a photon of light rather than
by the energy from a tungsten filament. Substances of this type

may be excited by ultraviolet light and emit visible light upon de-excitation. The process of using ultraviolet light waves to produce the excitation is called fluorescence. Light emitted by a fluorsecent lamp is produced in this manner.

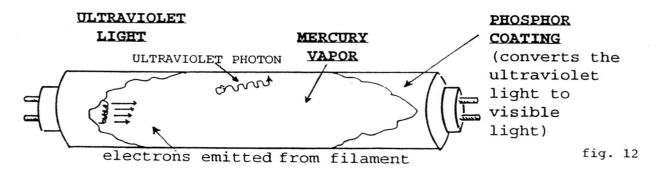

ULTRAVIOLET LIGHT   MERCURY VAPOR   PHOSPHOR COATING (converts the ultraviolet light to visible light)

ULTRAVIOLET PHOTON

electrons emitted from filament

fig. 12

In the glass tubing of a fluorescent light bulb the excitation of the gas by the filament produces ultraviolet light which has a high frequency and is not in the visible portion of the spectrum. The ultraviolet light excites the electrons in the thin chemical coating of phosphors on the inner surface of the glass tube. The phosphors are excited by the ultraviolet light in the tube and upon their de-excitation give off lower frequency wavelengths which correspond to the visible portion of the spectrum producing white light. Phosphors are a composition of matter which radiates visible light upon impact of light of a different wave length, usually ultraviolet. Thus, fluorescence is a process which uses ultraviolet waves to produce the waves in the visible portion of the spectrum (fig.12). Some fluorescent material appears to be invisible under normal light but can be seen under an ultraviolet lamp.

PHOSPHORESCENCE

Imagine going through a door which has a spring to make it close automatically behind you and having it get stuck in the open position. Then five minutes later it slams shut. This is the theory behind the process called phosphorescence. When excited, the electrons in the phosphorus material remain in a state of excitement because they get stuck in their higher orbits. There is a time delay between excitation and de-excitation. The electrons in this material begin to de-excite at a later time. When clock numbers and dials are coated with this phosphorus material, the electrons become excited by any visible light and they remain excited for several hours before they begin to de-

excite. If the lights are turned off or it gets dark outside,
the clock numbers and dials will glow in the dark as a result of
this delayed de-excitation.

LASER

Light Amplification by Stimulated Emission of Radiation

INCOHERENT LIGHT is light which is emitted from an incandescent
bulb. It contains many frequencies and wavelengths that are out
of phase (fig.13).

fig. 13

COHERENT LIGHT is light in which all waves have the same
wavelength, have the same frequency, are in the same phase and
move in the same direction (fig.14).

fig. 14

A LASER is an instrument that produces a beam of coherent light.
The laser is not a source of energy. It is a converter of
energy.  *Light amplification by Stimulated Emision of Radiation = Laser*

Incoherent light can be compared to people exiting a New York
City subway car. The people walk in all directions. Some take
big steps, some small steps, some walk fast and some walk slow.
While coherent light is analogous to a high school marching band
where all members of the band take the same size steps, with the
same speed and at the same time (fig.15).

fig. 15

Uses of the laser include eye surgery, welding, printing, global
surveying, communication and much more.

CONCEPT OF LENGTH

In Chapter I, it was noted that the meter was internationally redefined in 1967 as 1,650,763.73 wavelengths of the orange-red light emitted by the glowing krypton-86 gas.

1 meter = 1,650,763.73 wavelengths

Now after the discussion of wavelengths in this chapter, this method of defining the meter should be more meaningful. However, it was also pointed out that this method was changed in 1983 and that the meter was redefined as the distance through which light travels in one 299,792,458th of a second. It should be noted that the length of the meter has never changed. It is only the method by which we arrive at the meter that has changed.

EXAMPLES FOR THOUGHT XI

1. A characteristic common to sound waves and light waves is that they
    (1) are longitudinal   (2) are transverse
    (3) transfer energy   (4) travel in a vacuum

2. The Sun is approximately 93,000,000 miles away. Show that it takes approximately 8 minutes for sunlight to reach us. Since the distance was given in miles, you must use the speed of light as 186,000 miles/sec. and not 3×10^8 meters/sec. which is the same amount but in the metric system.

MATH HUMOR XI

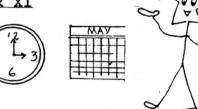

1. What occurs once in a second,
    once in a month,
    once in a century,
    not at all in a year,
    not at all in a week?

2. Find 111111111^2

3. What does HIJKLMNO represent?

4. Which of the following is correct?
    (1) 10^0 is less than 1^{10}
    (2) 10^0 is equal to 1^{10}
    (3) 10^0 is greater than 1^{10}

Chapter XII

Imagine making popcorn by pouring kernels into the bottom of a popcorn machine, adding heat and watching each kernel jump up as it pops. Without the heat, the kernels would remain dormant covering the bottom of the popcorn machine like a blanket. The same thing happens to the Earth's atmosphere. Radiant energy from the Sun is the major source of heat for the molecules in the Earth's atmosphere. This energy keeps the molocules in constant motion as opposed to being pulled towards the Earth's surface by gravity and just lying there covering the Earth.

The density of the atmosphere is greatest closest to the surface of the Earth and decreases with increasing altitude which means that the atmosphere is depth dependent. The following diagram gives the major classifications of the Earth's atmosphere.

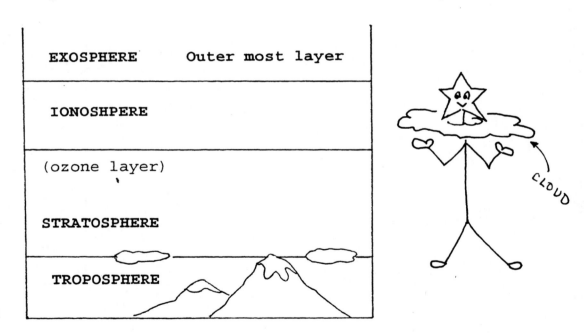

fig. 1

TROPOSPHERE:

Most of the air, moisture and dust of the atmoshpere is in the troposphere. This is where we live. It is the level in which there are clouds and from which we get our weather. Winds pick up water vapor from which clouds and rain form. The temperature drops rapidly with altitude in the troposphere. The air is warmest near the Earth's surface because the Sun's radiant wave energy heats the Earth which in turn warms the air near it.

STRATOSPHERE:

This layer extends above the troposphere. The atmosphere continues to get colder. At approximately the middle of the stratosphere the air begins to warm. This is because of the ozone layer. The ozone layer is a band of air which absorbs the Sun's ultraviolet rays and prevents most of the ultraviolet rays from reaching the Earth. This causes the atmosphere in the upper parts of the stratosphere to warm.

IONOSPHERE:

This layer extends above the stratosphere and receives its name from the ions that are found there. Ions are electrically charged molecules which are formed when these molecules are hit by radiation from the Sun. Since the ionosphere consists of these electrically charged particles, this layer has the ability to bounce low frequency radio waves back to the Earth. Without the ionosphere, these waves would continue into space because waves cannot bend to travel in a curve. Hot gases such as gases in the ionosphere which are composed of electrically charged particles are said to be in a plasma state. This is the fourth state of matter of which the other three are liquid, solid and gas. This plasma is not the same as the part of the blood called plasma.

EXOSPHERE:

This layer is the outermost layer. It is the space in which all the planets travel around the Sun.

AIR PRESSURE

The air molecules in the atmosphere are constantly being pulled down by gravity. The molecules closer to the Earth's surface are pushed even closer by the weight of other molecules above them. This push of air molecules against each other is called air pressure. In 1654, a famous experiment was conducted in the city called Magdeburg. In the experiment, two large hemispheres which became known as the Magdeburg hemispheres were placed together to form a hollow sphere. A team of horses was tied to the handle on one side and a second team of horses was tied to the handle on

the other side. The hemispheres were join so that they were air
tight and the air inside was syphoned out. With no air inside
the two teams of horses could not pull the hemispheres apart.
When the valve was opened, the air rushed back in and the horses
could now pull the hemispheres apart. From this experiment it
was concluded that when there was a vacuum (the absence of air)
inside the hollow sphere, the air outside presented a significant
force. This would mean that air has pressure. Since the density
of air decreases with altitiude, air pressure will also change
with altitude. Atmospheric pressure at sea level is:

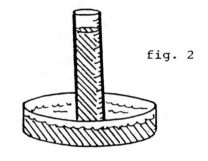

$$10^5 \text{ N/m}^2$$

In British units this is: 14.7lb/in^2

We are all subject to the pressure of the air around us. The
barometer is the instrument used to measure atmospheric pressure.
The first barometer was introduced by an Italian physicist named
Torricelli. He took a long vertical
glass tube that was closed at one end,
filled it with liquid, and inverted
it in a glass dish (fig. 2). The   fig. 2
closed end is now at the top. Some
liquid emptied into the dish leaving
a vacuum at the upper part of the tube.
This empty space was separated from the
atmosphere and could not exert a force
against the column of liquid. Thus, the atmospheric pressure on
the surface of the liquid in the dish is able to force the liquid
up the tube without any resistance. As a result the column of
liquid could be used to measure the pressure of the atmosphere.
Torricelli used water in the column. The use of water was
replaced by mercury because mercury is more dense and requires a
shorter column. An aneroid barometer also measures air pressure
but does not use liquid.

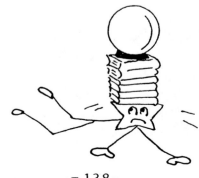

When you drink soda through a
straw, you are not sucking the
soda up through the straw. You
are reducing the pressure in the
straw allowing the atmospheric air
pressure on the liquid's surface
to push the soda up into the straw.
Sometimes if you sip slowly the
air pressure collapses the straw
before pushing the soda into the
straw (fig. 3).

fig. 3

All gases and liquids are fluids.
A fluid is defined as any substance
that flows readily. Fluid pressure
acts in all directions. If a boot
with holes in it is filled with water,
the pressure will push the water out
in all directions independent of the
orientation. The speed is the same
for all holes that are at the same
level because the pressure depends on
the depth not on the direction (fig. 4).

fig. 4

The pressure at a depth of one foot in a lake is the same as the
pressure at a depth of one foot in the water in your swimming
pool. Pressure at any point is exerted equally in all
directions. Imagine a thin small portion of water at any depth.
That portion of water is not moving which means that the force
pushing it up must be equal to the force pushing it down and the
force pushing it to the right is equal to the force pushing it to
the left.

fig. 5

Air is real and it takes up space. Try turning
a cup upside down and forcing it straight down
into a bucket of water (fig. 5). It will not
go down because something is already in the cup
(air). Air expands indefinitely, filling all the
space available to it and taking the shape of the
container. Gas is used in balloons because it
fills the balloon and prevents the atmosphere from collapsing

the balloon. Balloons filled with
helium rise because the weight of
the helium combined with the balloon's
weight is less than the weight of the
displaced air. Unlike water, there is
no sharp surface at the top of the
atmosphere. The exact height of the
atmosphere has no real meaning, but the

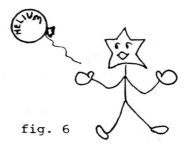

fig. 6

atmosphere gets thinner and thinner with increasing altitude. A
balloon filled with helium will rise as long as it displaces an
amount of air whose weight is greater than that of the balloon.
As the air becomes less dense with altitude and as the balloon
rises, a lesser weight of air is displaced by the balloon. When
the displaced amount of air equals the weight of the balloon, the
balloon will float. If it continues to rise, the pressure in the
balloon will become greater than the pressure outside the balloon
and the balloon will burst. This concept of the buoyancy of the
balloon is described by a principle introduced by Archimedes
which states that:

Air's mass:

$1\frac{1}{4} kg/m^3$

**"An object surrounded by a fluid is buoyed up
by a force equal to the weight of the displaced
fluid."**

In other words, if an object is lighter than the air or water
it displaces then it will float. If it is heavier than the air
or water it displaces, it will fall or sink.

A very famous story allegedly describes how Archimedes made the
discovery about his buoyancy principle. Archimedes was a Greek
philosopher who lived in the city of Syracuse in Sicily, which is
an island off the southern part of Italy. King Hero of Syracuse
gave some gold to a local goldsmith so that the goldsmith could
make a very ornate crown for the king. The crown was made and it
was more beautiful than the king had expected. But, somehow, the
king suspected that the goldsmith may have kept some gold and
substituted some other metal. Just weighing the crown would not
have proven anything because enough of another metal could have
been used to maintain the same weight. The king did not want to
melt down the crown, so he called upon Archimedes. He wanted
Archimedes to keep the crown intact and to figure out whether or
not some of the gold was replaced. Archimedes was accustomed to
going to public baths and one day as he was about to go into the

water, the idea came to him on how to test the crown. The story has it that at that moment he ran down the street yelling, "Eureka, Eureka", which means, "I have found it. I have found it."

What Archimedes found was that when he submerged his foot beneath the water level, the level of the water in the bathtub rose. As he put his foot into the water, some of the water was pushed out of the way. In other words, the water was displaced by his foot. The total volume displaced was the volume of his foot.

Using this concept Archimedes found that the volume of the crown was greater than it should have been and therefore, he concluded that the goldsmith stole some of the gold. If the goldsmith took out some of the gold he had to add enough of another metal to equal the weight of the missing gold. If the new metal were lighter than gold then more metal would be needed to make up the equal weight and hence the crown would have a greater volume. When placed in water, the greater volume would displace more water. Archimedes knew that objects of the same weight do not necessarily have the same volume. He took a piece of pure gold which had the same weight as the weight of the crown. If the crown were made of pure gold, the crown and the piece of pure gold would each displace the same amount of water but they did not. The crown spilled over a larger volume of water. By this test Archimedes concluded that the gold in the crown was impure.

TRY THIS EXPERIMENT:

ROCK   fig. 7

<u>A CLASSIC QUESTION</u>

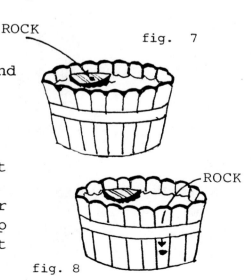

Suppose a bucket was filled with water and in it was a toy rowboat. A small heavy rock was placed in the boat (fig. 7). The rock pushed the boat lower into the water and raised the water level. Now suppose the rock is taken out of the boat and dropped into the water in the bucket with the boat still floating in the water (fig. 8). Will the water level now go up or go down if the rock is not in the boat but just in the water?

ROCK

fig. 8

Answer: With the rock in the boat the boat becomes heavier and displaces more water. The rock in the water without the boat displaces very little water. Its volume is very small. When the rock is taken out of the boat the boat rises and the water level goes down. The rock alone in the water does not displace much water so the net result is that the water falls.

BERNOULLI

Until this point the discussion in the chapter referred to fluids that were not in motion. With motion, there are other influences to consider. One of the principles of a moving fluid was introduced in the 18th century by a Swiss scientist, named Daniel Bernoulli. Bernoulli's principle states that the pressure in a fluid decreases with increased velocity of the fluid. A fluid speeds up in a constricted region if flow is to be continuous. Bernoulli reasoned that the extra speed was acquired at the expense of a lowered internal pressure.

Bernoulli's principle states that:

> **"In a flowing liquid or gas the pressure is the least where the speed is the greatest."**

This can be explained with the following illustration.

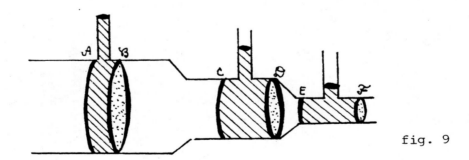

fig. 9

The water in the vertical columns stand highest in the widest part showing the pressure is the greatest where the water flows slowest (fig. 9). How do we know where the water is flowing the slowest? To explain this, we should concentrate on the speed of any one given molecule. Since the water is flowing continuously, then a molecule must get from point A to point B in the same time that a molecule must get from point C to point D and in the same time that a molecule must get from point E to point F. But the distance betwen points E and F is greater than the distance between points C and D which is greater than the distance between points A and B. This means that a molecule must move faster in the constricted region with the greater distance in order for the water to flow continuously. This is where the pressure is the least. The kinetic energy increases with speed and the potenetial energy or pressure decreases because of conservation of energy.

APPLICATIONS OF BERNOULLI'S PRINCIPLE

If you are traveling at 60 mph on a very calm day and you place your hand out the window, you will feel a wind blowing above and below you hand at 60mph. The car has generated a relative wind that is moving at the same speed but in the opposite direction. The same is true for an airplane. Now suppose there are two air molecules at the lead edge of the wing. Suppose that one molecule goes above the wing and the other goes below the wing. These two molecules will meet again at the back edge of the wing if the air is flowing smoothly. This is a fact of nature.

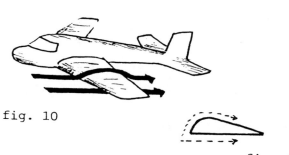

fig. 10

fig. 11

Since the upper surface has a curve to it, the molecule which travels on top must go a greater distance. (The shortest distance between two points is a straight line.) If the top molecule has to go a greater distance in the same time interval then it must travel faster. All the air on top moves faster for this reason and according to Bernoulli where the air moves faster the pressure is the least. The reduced pressure on the top means an upward force. This is one factor that produces the lift for the plane. Other factors that affect the lift are

...the angle of attack
...the wing area
...air density
...the plane's speed

TRY THIS EXPERIMENT:

How fast the plane is flying will
determine how fast the air flows
across the surface. You could
experiment with these factors by
trying the following simple experiment.
Hold a piece of looseleaf paper by the
edge so that it is curving downward while
you are standing still. Then run with
the paper (fig.12). Notice that the free
end of the paper is lifted up. You might
try holding the edge of a sheet of paper
in front of your mouth and blowing across
the top surface. Notice again that the
paper's free end rises (fig.13).

fig. 12

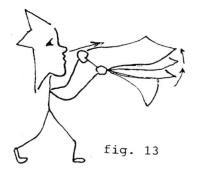

fig. 13

Another time when Bernoulli's principle is at work is when water,
running through a hose, is made to speed up. Cover a slight part
of the nozzle. The water going through the constriction has to
speed up.

An atomizer at the top of a bottle of perfume
works on the basis of Bernoulli's principle.
Pinching the bulb causes a flow of air across
the top of the vertical tubing which in turn
reduces the air pressure. The pressure on the
surface of the fluid is now greater and the
fluid rises in the vertical column (fig.14).

fig. 14

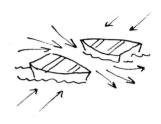

fig. 16

When water flows faster between two ships, the ships will move closer sideways. This again uses Bernoulli's principle where the fluid moves faster the pressure is the least. With less pressure between the boats, the the pressure on the ouside pushes them together (fig.16). For the same reason a solid cement dock will do damage to a ship tied to the dock because water moving quickly between will cause a reduced area of pressure and the ship will slam sideways into the cement wall. This is why docks are built out of poles which allows the water to flow evenly and at the same speed on all sides (fig.17). Bernoulli's principle helps us understand why cars tend to move sideways

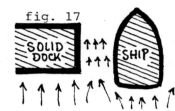

fig. 17

towards a passing truck on the highway. The air is moving fastest on the side of the car nearest the truck, resulting in a deacrease in pressure on that side. The higher pressure on the outer side pushes the car toward the truck.

TRY THIS EXPERIMENT:

Place a paper over two books and blow in the space between the books. Notice the paper will sink because the air pressure is less below the paper (fig.18).

fig. 18

THREE MORE EXAMPLES OF BERNOULLI'S PRINCIPLE

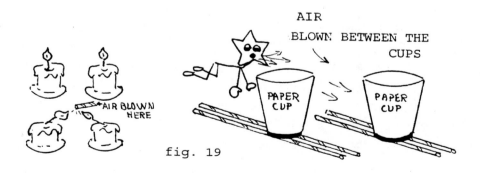

AIR BLOWN BETWEEN THE CUPS

fig. 19

The vacuum cleaner has been incorrectly named because there is no vacuum. Instead it is Bernoulli's principle at work. Through the motion of air, the pressure inside is decreased. Because the air pressure on the outside is greater it moves to the lower pressure area and any litter in the way moves with it. 'High pressure wants to become low pressure' is a concept which might help describe what is happening.

TORNADO WINDS

There is a belief that if a tornado is headed in the direction of a house, the windows should be opened a bit to save the house from total destruction. The theory is that this would make the pressure outside and inside equal. The moving air outside the house creates low pressure with respect to the stationary air in the house. This might lead to the idea that a house in the path of a tornado explodes because of the greater inside pressure. The fact that the house could be saved by opening the windows is actually not correct. In reality the buldings are destroyed from the outside by the force of the winds, even though it looks like an explosion from inside. Opening windows will not help against these rapidly moving winds.

EXAMPLES FOR THOUGHT XII

1. Using a piece of clay and a bowl of water, try to get the clay to float. It could be molded into any shape.

MATH HUMOR XII

1. What did one physics book say to another physics book?

2. Did you hear about the person who finally figured out that an A.M. radio could be heard in the afternoon too.

3. Using a calculator: Enter 26
   Multiply by one million
   Add 522,959
   Mutliply by 2

   Turn the caluculator upside down.
   What do you see?

4. Given the following arrangement of numbers:
        8,5,4,9,1,7,?,?,?,?
    What are the next four numbers after the "7"?

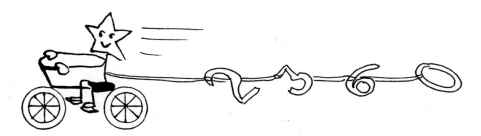

5. Think of the number of the month
    in which you were born.
    Multiply that number by 2.
    Add 5.
    Multiply by 50.
    Add your age.
    Subtract 365.
    Add 115.
    What do you notice about your result?

6. Without looking at your watch, try to recall if your watch
    has numbers or if it has markings to indicate the hours
    and minutes. If it has numbers, how many are there? Most
    individuals will not know because many times we look but
    we do not see.

Atmosphere pressure is $10^5 N/m^2$ at sea level, this means that

✓a. the weight of a column of air/sq meter in cross-section
    extending up to the top of the atmosphere is $10^5 N$

b. the weight of a cubic meter of air at sea level is
    about $10^5 N$

c. the desity of air is $10^5 N/m^3$

d. None of the above.

Chapter XIII

ENERGY AND ENERGY RESOURCES

The average person uses many energy servants each day. The
following is a list of only ten of these energy servants.

    1. waking up in the morning to an electric clock radio
    2. turning on the lights
    3. taking a shower with hot water
    4. using an electric hair dryer
    5. making coffee or hot cereal, or
       using a toaster
    6. taking a bus, train or car to
       school or work
    7. listening to a radio or watching
       television
    8. heating or cooling the home or office
    9. using a telephone
  10. using a computer

The rate at which the U.S. uses energy and will continue to use
energy is growing and will continue to grow as we 'enter the
21st century. Our non-renewable resources will not keep up with
this pace. Over the years the supply of non-renewable energy has
been supplemented with other forms of energy.

Our major energy resources are oil, natural gas, coal,
hydropower, nuclear power, wind power, solar power and others.
We use these resources in three major ways:

    1. commerical and residential purposes (stores and homes)
    2. industrial purposes (factories)
    3. transportation (cars, buses, trains and planes)

FOSSIL FUELS

A good percentage of the energy that the U.S. needs is filled by
the use of fossil fuels which consists of COAL, CRUDE OIL, and

NATURAL GAS. They are called fossil fuels because they are the remains of plants and animals that lived a long time ago.

COAL:

Coal is a fossil fuel which is obtained from mines. One negative feature in the use of coal as a source of energy is that when it is burned, it produces sulfer oxides when combined with water vapor produces sulfuric acid. Winds in the atmosphere carry this acid to other regions where it comes down with the rain as acid rain. This produces problems to the environment by increasing the acid levels in the soil and water. Crop production could be damaged by too much acid. Smoke from coal mines also causes air pollution.

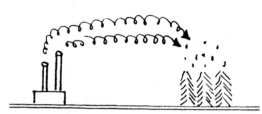

CRUDE OIL:

Crude oil is a mixture of hydrocarbons—hydrogen and carbon. Most of our supply of crude oil which is another name for petroleum comes from wells in the U.S. and some is imported from foreign countries. Burning gasoline and other fuels obtained from crude oil causes air pollution. Petroleum which is pumped out of the ground is usually a black sticky mixture that is quite useless until it has been refined. Refining is the process of separating the petroleum mixture into its various parts. Boil water and you get steam. Boil oil and you get everything from kerosene and gasoline to asphalt. These and other products are distilled from crude oil. Butane and propane boil off right away. Asphalt boils off last. Gasoline, kerosene, diesel fuel, and fuel oil are in between. Boiling petroleum separates each of the compounds at different temperatures and the process is called distillation.

NATURAL GAS:

The third fossil fuel is natural gas which is used for cooking and heating. It is a clean fuel and produces little or no air pollution. METHANE, another name for natural gas, can be produced in nature by the bacterial decay of vegetation and animal wastes in the absence of air. The amount of garbage produced in the U. S. is staggering. It contains much animal wastes and other forms of biomass. Biomass refers to plants, animal wastes and all other organic matter that can be used as a source of energy. One way to release energy from solid wastes is to burn them. They must be processed before the energy from the waste can be used to produce steam for an electric generator. Some of what we grow is used for food, paper, building material and some goes to waste. Agricultural wastes could be peanut shells and corn husks. There are ways to make use of this form of biomass which would otherwise go to waste. It is also a renewable resource. Methane does not have to come from waste alone. It also could be farmed just for energy purposes. Alcohol could be made from wood chips, corn, sugar cane, beets and other plants. There is a process using, for example, corn, which produces ethanol from which pure alcohol is removed. It is then mixed with gasoline to produce gasohol and is referred to as grain alcohol. Although alcohol in the form of ethanol or methanol can be used as part of an automobile's fuel and although it has many advantages, it may not be the most economical or efficient way to use biomass material.

OTHER ENERGY SOURCES

GEOTHERMAL POWER:

Inside the Earth is a huge amount of heat in areas where steam or very hot water exists. Heat within the Earth is called GEOTHERMAL ENERGY and the temperature of this heat increases as one goes deeper and deeper into the Earth. It is necessary to drill into these spots to pump out the energy which could be used to power up a generator which in turn could be used to produce electricity. One of the problems associated with this source of energy is finding a method of preventing the drill from overheating. Another problem is that once the source of energy is used up, reheating is very slow. Also, geothermal wells are more expensive to build than oil wells. The diameter of the wells is much larger and they are drilled through harder rock. Another fact is that geothermal wells require the power plant to be built at the site of the well because hot water or dry steam which is the essence of geothermal power cools and loses pressure if moved too far. Geothermal energy must be converted to power on the site and the electricity must be moved in expensive power lines to the point of use. In contrast, oil from remote fields can be shipped to a refinery and then on to where it is needed. However, geothermal thermal energy is, for the most part, pollution free and safe which makes it environmentally desireable. Power plants producing this type of energy have been in operation in Italy and California for many years and are functioning very successfully.

HYDROELECTRIC POWER:

The rush of a waterfall is another way of producing energy for electricity. Potential energy is changed to kinetic energy. The power of the waterfall is used to cause the blades of a turbine to spin and thus power up a generator which, in turn, produces electricity. Hydroelectric power is desireable in that it produces little, if any, pollution and it is undesireable in that there are not many good locations. There are also high costs involved and esthetically a beautiful waterfall is nicer to look at than a hydroelectric power plant. In addition, another problem is that building a dam or changing the flow of a river can harm the environment by killing marine life and destroying plant life.

SOLAR POWER:

Solar energy consists of radiation from the Sun. The technology for converting this radiant solar energy into electric power is available but some problems exist. Solar energy systems need a method of storing the energy for the time when there is no sunlight. One way is to heat either water, rocks or gravel. Another method is to use photoelectric cells.
Some of the advantages of using solar power are as follows:

        ...there is a limitless supply
        ...it produces no air pollution
        ...it produces no water pollution
        ...it produces no thermal pollution
        ...it produces no harmful wastes
        ...there is no possibility of an explosion
        ...it conserves the Earth's resources
        ...the technology is available
Some of the disadvantages of using solar power are as follows:
        ...it is expensive
        ...it may not be practical to have one major collecting
            source
        ...it may be more econonmical to have each home

responsible for its own solar panels and its own
solar energy

WIND POWER:

Wind is used to drive ships, pump water, to grind grains and to
create electricity. Windmills have been built to make use of the
power of wind and are capable of producing electricity. As with
solar power, wind power could be harnessed at large central
stations or as small sized home units. The electricity which the
windmills generate could be fed directly to the user. A backup
power source would be needed
for the times when the wind would not supply enough
power to meet the users' needs. If a large enough
wind generating system were built, there would
always be a sufficient amount of electricity
produced by the system to meet the users' needs.
The greatest negative feature is that it is very
expensive.

NUCLEAR ENERGY

NUCLEAR FISSION:

Nuclear fission occurs when the nucleus of a uranium or
plutonium atom is split by a free neutron causing the release of
energy. The split produces two differnt nuclei called fission
fragments as well as several free neutrons and an enormous amount
of energy. The total mass of the fission fragments and the
neutrons is less than the mass of the original uranium atom and
the neutron The missing mass is converted into energy. The free
neutrons can cause the splitting of other uranium nuclei. An
uncontrolled succession of splitting could produce a chain
reaction which would result in an explosion (fig. 1).

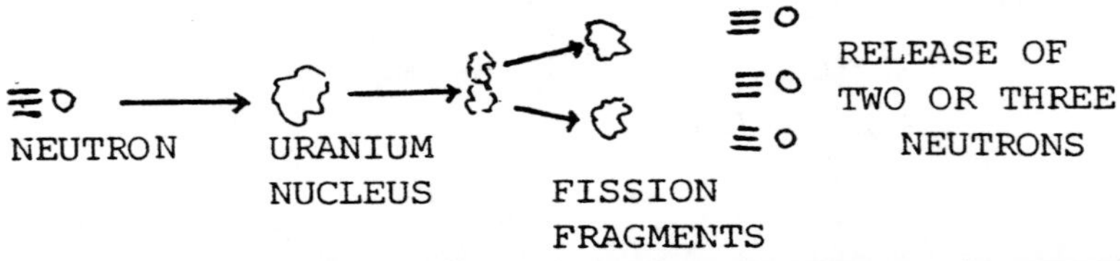

fig. 1 NEUTRON URANIUM
NUCLEUS FISSION
FRAGMENTS RELEASE OF
TWO OR THREE
NEUTRONS

One of the functions of a nuclear reactor is to control the
fissioning by controlling the number of available neutrons. This
is accomplished by neutron absorbing rods, also called
controlling rods, which consists of elements that absorb the
extra free neutrons. In other words, the release of energy can
be controlled and kept at a steady or constant rate. Control

rods are placed between the fuel rods so as to control the number of available neutrons for inducing fission and for preventing a chain reaction. The control rods can be adjusted according to the amount of energy that is required. Pushing the rods all the way into the reactor stops the fission reactions (fig.2).

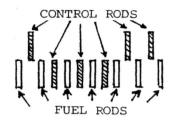

fig. 2

The coolant in a reactor removes heat energy from the reactor by being pumped through the fuel rods and absorbing the heat released by the fissioning process. Fuel rods are tubes in the reactor's core containing the fissionable fuel. The coolant which could be water carries the heat around other pipes containing pressurized water in the heat exchanger. The heat from the coolant changes the water in the pipes to steam. Steam carries the heat away from the reactor to a turbine. Thus, a nuclear reactor producess steam which is used to do work consisting of rotating the blades of a turbine. This powers up a generator and in turn produces electricity. A huge amount of electric power is generated with a relatively small amount of fuel.

Water flowing through the fuel rods is not only a coolant but also acts as a moderator. Slow neutrons are needed for the fission process but the neutrons which are released from the fission are fast moving neutrons. These fast moving neutrons are slowed down by colliding with the nuclei in the moderator where they lose some of their energy. Hydrogen atoms in water are very effective in slowing down these fast neutrons. There are some materials other than hydrogen that are used as moderators in some reactors.

The fissionable fuel in the fuel rods does get exhausted and these spent fuel rods must be replaced. The spent fuel rods contain the fission fragments which are radioactive. This radioactive waste must be stored safely for many years because the half-life is very long. It may take thousands of years before the radioactive waste decays and is no longer radioactive.

One of the major problems inherent in using nuclear power as a source of energy is that the by-product consists of radioactive wastes. As mentioned, this radioactive waste material must be stored safely for thousands of years until the radiation decays. Underground storage requires careful monitoring because accidental escape of radioactive wastes could kill plant and animal life. In addition, provisions must be made to ensure that hot water, produced in the process of creating nuclear energy, is not dumped into rivers or oceans where the heat could harm marine life.

A major concern about nuclear reactors is the loss of coolant causing a buildup of heat in the core. If the reactor loses its water, as happened at Three Mile Island in Pennsylvania in 1979, the reactor can become dangerously overheated possibly causing a meltdown. Because of a series of errors, over half of the core melted and the reactor was ruined. The construction of the containment building helped to confine the accident to the nuclear plant. Another major concern arose in 1986 when an explosion occured at the nuclear power plant in Chernobyl in the U.S.S.R. (as it was known then). This nuclear accident poured tons of radioactive material into the atmosphere.

Although nuclear reactors are used throughout the world for many peaceful purposes there are many people who are opposed to the use of this form of energy. The following is a list of some of the pros and cons.

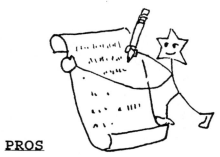

	<u>PROS</u>	<u>CONS</u>
EFFECTS ON THE ENVIRONMENT	Nuclear power is the cleanest fuel	Not if there is an accident
POSSIBILITY OF A SERIOUS ACCIDENT	Small	With more plants the possibility rises
RADIATION LEAK	No measureable effect	Any release is unacceptable
THERMAL POLLUTION	Not bad	Heat into the lakes upsets the ecology
STORAGE OF WASTES	Can be worked out	Too many unanswered questions
NEED FOR MORE POWER	Needed to save fossil fuels and energy Needed because consumption is growing at a rapid rate	Conservation and alternative energy sources should be used
LARGE CENTRAL GENERATING PLANTS	Very efficient	Should be collected individually by user
COST	Nuclear power is relatively cheap	The cost of radioactive waste disposal must be figured also
POLITICAL SECURITY OF FUEL SUPPLY	Uranium supply is politically secure	Only the sun and wind are politically secure

BREEDER REACTORS:

A breeder reactor produces both energy and nuclear fuel. It produces more fuel than it uses by converting non-fissionable nuclei into fissionable nuclei. It actually expands the supply of fissionable material. It can provide enough nuclear fuel for its own operations and can provide additional fuel. In a breeder reactor, the moderator does not slow down fast neutrons as it does in an ordinary nuclear reactor. With this process it is possibe to extend the supply of fissionable material by converting a nonfuel into a fuel. An ordinary reactor uses fissionable fuel such as uranium-235 to produce energy. A breeder reactor is set to use a fissionable fuel to produce energy and is also set to convert a non-fissionable fuel such as uranium-238 into a fissionable fuel such as plutonium-239. The result is more fissionable fuel which could be used to produce more nuclear energy.

THE FUNCTION OF ANY POWER PLANT:

The function is to drive a turbine to power generator which creates electricity. The only difference among power plants are the sources of energy needed to drive the turbine. Once high pressure steam is created a nuclear power plant and a coal power plant are the same.

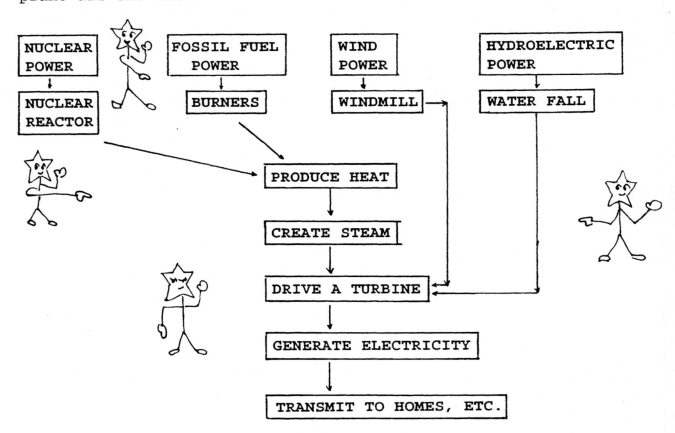

THERMONUCLEAR POWER

FUSION:

Fusion occurs when two nuclei are blended together to form one nucleus with the release of a tremendous amount of energy. It is the opposite of nuclear fission which requires splitting in order to produce energy. In an atom the nucleus is positively charged and the electron is negatively charged. The positive charge of the nucleus attracts the opposite negative charge of the electron and binds the electron to the atom. Opposite electrical charges attract and similar electrical charges repel each other. It is this opposing force between positively charged nuclei that makes fusion a very difficult process. Hydrogen can undergo fusion but tremendous temperatures must be reached for fusion to occur. The Sun has the temperature needed for fusion to take place. Through a complex series of steps the Sun's hydrogen is blended to form helium and at the same time this process releases a huge amount of radiant energy. Fusion produces the radiant energy in the Sun and other stars as well. This is a thermonuclear reaction because the reaction is the result of enormously high temperatures or thermal energy. High temperature matter is the fourth state of matter called plasma. When a liquid is heated sufficiently, it becomes a gas. When a gas is heated further, it becomes a plasma.

If small scale fusion reations could be produced in a laboratory, it would require more energy to cause the reaction than resulted from the reaction. On the positive side, if it could be done there would be no problem of radioactive wastes resulting from the fusion. Experiments are being conducted to see if there are ways to cause fusion to occur without using these very high temperatures. This type of fusion is called cold fusion. In March, 1989, two scientists said that they caused a fusion reaction at room temperature. Unfortunately, this procedure was found to be ineffective and research in this area of cold fusion is still continuing around the world.

A NEED FOR CONCERN

There is an old science experiment in which an animal was placed in water at room temperature. Gradually more water was added and the temperature of the additional water was made higher and higher bringing the water with the animal in it to the boiling point. The animal stayed in the water and did not react to the boiling water. Another animal was added to the boiling water and immediately jumped out with a look of pain. A change that occurs gradually most often goes unnoticed. This scenario of the animal in the water can be compared to the depletion of our non-renewable energy resources which might not be felt until it is too late. Even though there seems to be a sufficient amount for our needs now, it may not last if our needs grow exponentially. The purpose of the following story is to illustrate exponential growth and to draw attention to the need for concern.

Suppose that a rare animal lives on a secluded island in a make-believe world that has only three islands and the rest of this make-believe world is water. This animal lives only on land and reproduces by doubling. The doubling time is one month. This means that at the end of one month, there are two such animals on this island in this make-believe world. At the end of two months, there will be four animals, and so on. Each island is very small and in ten years the first island is completely covered with these rare animals. There was no more space for another animal to inhabit this first island. Luckily or not so luckily, there was one animal who saw the problem in the fifth year, when the island was only half covered with animals. He left the island in search of more space and found that this make-believe world had two more similar islands. He came back feeling good and thinking they had nothing to worry about because there was plenty of space out there. The next five years flew by and the first island was saturated with animals. At the same rate of doubling it took one month and the second island was completely saturated with animals, too. Now there was only one uninhabited island left in this make-believe world. When these animals doubled in the next month they needed two more islands, but there was only one island left. At this point it was too late to plan.

This is a very simple story that makes a very important point. It is that our need for energy is growing at a rapid pace and unless we conserve or find alternative energy sources we will run out of our non-renewable souces faster than we think.

This type of growth is called exponential growth. There is another interesting story which shows how exponential growth works. There was a ruler who was pleased with the game of chess that a subject in his kingdom invented for him. The ruler wanted to reward this subject and asked what the subject wanted. His request seemed modest so the ruler honored it. The request was that the subject be given one grain of wheat on the first square of the chess board, two grains on the second, four on the third, and so on, doubling it for each successive square. When he finished the ruler owed the subject all the wheat in the kingdom plus more. The following table explains the problem.

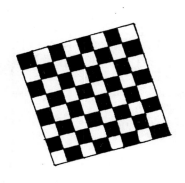

SQUARE NUMBER	NUMBER OF GRAINS ON EACH SQUARE	TOTAL GRAINS
1	$1 = 2^0$	$1 = 2^1 - 1$
2	$2 = 2^1$	$3 = 2^2 - 1$
3	$4 = 2^2$	$7 = 2^3 - 1$
4	$8 = 2^3$	$15 = 2^4 - 1$
5	$16 = 2^4$	$31 = 2^5 - 1$
6	$32 = 2^5$	$63 = 2^6 - 1$
7	$64 = 2^6$	$127 = 2^7 - 1$
$\vdots$	$\vdots$	
64	2^{63}	$2^{64} - 1$

$$2^{63} = 9,223,372,036,854,775,808$$

This is read nine quintillion 223 quadrillion 372 trillion 36 billion 854 million 775 thousand 808.

$$2^{64} - 1 = 18,446,744,073,709,551,615$$

Thousands of bushels of wheat would be needed to satisfiy the request. This would be much more wheat than the ruler had. It was even more than the world's entire production of wheat for several thousand years.

If this is not impressive, try calculating the amount of money that would accumulate if interest were compounded anually. For example, suppose $1000 were invested at 6% interest compounded annually. The $1000 would amount to $1000(1+.06) or $1060 at the end of the first year. This would be the principal for the second year's computation. At the end of the second year the amount would be $1000(1+.06)(1+.06).

This equals $1000(1+.06)^2$

At the end of 3 years it would be $1000(1+.06)^3 = \$1191.02$
At the end of t years it would be $1000(1+.06)^t$

In general, if P = principal
                r = rate of interest
                    compounded annually
                n = number of years

        then $A = P(1+r)^n$

Exponential growth produces greater quantities much faster than linear growth.

$1 invested from the year one to the year 1997 at 3% interest compounded annually would be approximately equal to :

$$A = \$1(1+.03)^{1997}$$
$$A = 4.324518273 \times 10^{25}$$
$$A = 43,245,182,730,000,000,000,000,000$$
Approximately $43,000,000,000,000,000,000,000,000

The same amount invested at the same rate for the same time period but with simple interest would equal $60.91.

        1997 X .03 = $59.91; $59.91 + $1.00 = $60.91

HOW MUCH SHOULD WE BE CONCERNED?

We will certainly have enough energy for many years
to come. However, if we don't plan now, there will
be future generations who will have a problem. There

is a story that sums up the meaning of this chapter.

The ruler of a small island conducted a race and said that a reward would be given to the best racer. The people asked, "What constitutes the best racer?" The ruler replied, "Prepare yourself as best as you know how and you will see." For weeks the people practiced and then the time came for the actual race. The first person to finish went up to the ruler and said, "I was the first, therefore, I am the best." The ruler replied, "You are good, but not the best." The second runner to finish went up to the ruler and said, "I may have come in second, but my racing style makes me the best." "No," said the ruler. In the meantime, many people completed the course. But one man while racing encountered a pile of rocks which the earlier runners just ran around to avoid. This man moved the rocks to the side, one at a time, so that they would not be in the way of the of the runners to follow. In doing this he found an envelope, containing money, under one of the rocks. He put the envelope in his pocket and continued to move all of the rocks out of the way. When he finished the race, he went up to the ruler and gave him the envelope with all of the money. He explained what he did and where he found the money. The ruler said, "That is the reward and it is yours. You took the time to clear the path and made it easier for those who followed you. This is what all of us should do for future generations."

The lesson to be learned from this chapter is summarized by the three "R's".

REDUCE
REUSE
RECYCLE

It is important to reduce our consumption. This means using less energy. Then we must learn to reuse items such as plastic bags and avoid being a "throw-away" society. Finally, recycling, which is being done in many communities, is also necessary to conserve our energy sources.

1. List as many of the energy servants that you use during a week.

2. Referring to the example in the text, compute the interest on one dollar for 1997 years compounded annually at 2%.

3. Referring to the example in the text, compute the interest on one dollar for 1997 years at 2% simple interest.

4. Which of the following options would you chose: all the money that starts with one cent and doubles every day for thirty days or one million dollars? Why?

5. Take any piece of paper and fold it in half as many times as you can. How many folds did you get?

6. A microbe doubles every second and in one minute fills a jar. Suppose the doubling starts with four microbes, how long will it take before the jar is filled?

7. Suppose there was a lake with one lilypad on it and suppose the number of lilypads doubles every 24 hours. If it takes 60 days for the lake to be covered, on what day is the lake one-quarter covered.?

MATH HUMOR XIII

1. Imagine that you are in a sealed room where the walls are made of cast iron. There are no windows and no doors. If there is only a piece of wood and a saw in the room, how would you get out?

2. What is H_2O?
    What is H_2O_2?
    What is H_2O_4?

3. Teacher: "Which is more important to us:
        the Moon or the Sun.?"
   Child: "The Moon."
   Teacher: "Why?"
   Child: "The Moon gives us light at night
            when we need it. The Sun only
            gives us light in the daytime
            when we don't need it."

4. If you had to swallow one pill every half hour, how much
   time would it take you to swallow three pills?

5. Divide twenty by one-half and add three. What is your
   answer?

Review

Chapter XIV

FINAL EXAM REVIEW

1. **ABSOLUTE ZERO:**
   THE LOWER LIMIT OF TEMPERATURE ON THE KELVIN SCALE AND
   THE POINT AT WHICH THERE IS NO THERMAL ENERGY. $-273C°$

2. **AMPLITUDE:**
   THE DISTANCE A TRANSVERSE WAVE RISES OR FALLS WITH RESPECT
   TO A BASE LINE.

3. **ARCHIMEDES' PRINCIPLE:**
   THE BUOYANT FORCE ACTING ON AN OBJECT IN A FLUID. THIS
   FORCE IS EQUAL TO THE WEIGHT OF THE FLUID THAT IS BEING
   DISPLACED BY THE OBJECT.

4. **ATMOSPHERE:**
   THE LAYER OF GASES SURROUNDING A PLANET SUCH AS THE EARTH.

   QUESTION:
4-1. The lowest region of the atmosphere is called the
   ✓(1) troposphere   (3) ionosphere
    (2) stratosphere   (4) exosphere

5. **ATMOSPHERIC PRESSURE:**
   THE PRESSURE EXERTED BY THE WEIGHT OF A COLUMN OF AIR.
   THE AIR PRESSURE OF THE EARTH'S ATMOSPHERE AT SEA LEVEL
   IS $100,000N/m^2$ OR $10^5N/m^2$.

   QUESTIONS:
5-1. When a person is sipping a drink through a straw,
   the liquid rises because of
    (1) Archimedes' Principle   ✓(3) atmospheric pressure
    (2) Bernoulli's Principle   (4) none of these

5-2. When a plunger adheres to a wall, it is
    (1) held there by a vacuum   (3) both (1) and (2)
   ✓(2) pushed by the atmosphere (4) neither (1) nor (2)

6.   **BERNOULLI'S PRINCIPLE:**
WHEN THE SPEED OF A MOVING FLUID INCREASES, THE PRESSURE
DECREASES.

QUESTIONS:
6-1. The Bernoulli principle explains why ships are drawn
toward each other sideways when the ships are close and
moving in
(1) the same direction (3)✓the same or opposite directions
(2) opposite directions (4) not enough information

6-2. Airplane flight is explained by _____principle.
(1) Archimede's (2) Pascal's (3) Boyle's (4) Bernoulli's
✓

7.   **CELSIUS:**
TEMPERATURE SCALE IN WHICH 0° IS THE FREEZING POINT AND
100° IS THE BOILING POINT.
ANDERS CELSIUS WAS A SWEDISH ASTRONOMER.

8.   **CHAIN REACTION:**
THE REACTION IN WHICH MANY NEUTRONS ARE RELEASED BY THE
FISSIONING PROCESS AND THIS CAUSES UNCONTROLLED FISSIONING.

9.   **COHERENT LIGHT:**
LIGHT WAVES WHICH HAVE THE SAME AMPLITUDES AND WHICH RISE
AND FALL TOGETHER.

10.   **CONDUCTION:**
THE METHOD BY WHICH HEAT TRAVELS THROUGH A SUBSTANCE BY
WAY OF COLLISIONS BETWEEN ADJOINING MOLECULES. THIS OCCURS
MAINLY IN SOLIDS.

QUESTION:
10-1. A good conductor of heat is a
(1) poor insulator   (3) neither (1) nor (2)
(2) good insulator   (4) both (1) and (2)

11.   **CONSERVATION OF ENERGY:**
ENERGY CANNOT BE CREATED NOR DESTROYED. IT MAY BE
TRANSFORMED FROM ONE FORM INTO ANOTHER FORM, BUT THE TOTAL
AMOUNT OF ENERGY NEVER CHANGES.

11-1. In the diagram at the right, a pendulum is released from point A and swings in the direction of point B. At which position does the bob have the maximum kinetic energy?

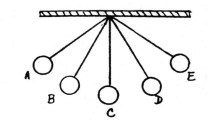

(1) A (2) B (3) C (4) D
√

11-2. Use the same diagram as in question 11-1. If air resistance and friction are ignored, the bob's kinetic energy at point A is the _____ as the potential energy at point B. (Note: B is half the distance between A and C.)
(1) half as great   √(3) the same
(2) four times as great   (4) not enough information

11-3. As an object falls freely near the surface of the Earth, the object loses gravitational potential energy. This is equal to the object's
(1) loss of mass   (3) loss of kinetic energy
(2) gain of mass   √(4) gain of kinetic energy

12. **CONSERVATION OF MECHANICAL ENERGY:**
THE TOTAL MECHANICAL ENERGY IS CONSTANT. THAT IS, THE SUM OF THE KINETIC ENERGY AND THE POTENTIAL ENERGY REMAINS THE SAME.

13. **CONTROL ROD:**
A NEUTRON ABSORBING ROD WHICH IS PLACED IN A NUCLEAR REACTOR IN ORDER TO CONTROL THE NUMBER OF AVAILABLE NEUTRONS.

14. **CONVECTION:**
THE METHOD BY WHICH THERMAL ENERGY TRAVELS THROUGH A FLUID. THIS OCCURS IN LIQUIDS AND GASES.

QUESTION:
14-1. The rising of warm air has to do with the fact that faster moving molecules tend to move to regions of less
(1) density (2) pressure (3) opposition (4) all of these
√

15. **DE-EXCITATION:**
THE PROCESS OF AN ELECTRON RETURNING TO ITS ORIGINAL ENERGY
LEVEL FROM A HIGHER ENERGY LEVEL AND IN SO DOING EMITS
RADIANT WAVE ENERGY WHICH IS PROPORTIONAL TO THE FREQUENCY
OF THE RADIANT WAVE.

16. **DOUBLING TIME:**
THE TIME NECESSARY FOR A GIVEN POPULATION TO DOUBLE AT
REGULAR INTERVALS. IT IS ALSO REFERRED TO AS EXPONENTIAL
GROWTH.

QUESTION:

16-1. If a lily pond started with a single lilipad and each
day the number doubled until it was completely covered
on the 30th day, then on the 29th day the pond was one
half covered. If instead the lily pond started with two
lily pads, on what day would the pond be completely
covered?
(1) day 15   (2) day 28   (3) day 29   (4) day 30

17. **EFFICIENCY:**
THE RATIO OF USEFUL WORK PRODUCED BY A MACHINE TO THE
TOTAL ENERGY PUT INTO RUNNING THE MACHINE. IF A MACHINE
IS 75% EFFICIENT, THEN THREE FOURTHS OF THE INPUT ENERGY
IS CONVERTED INTO A USABLE FORM AND ONE FOURTH BECOMES
UNUSABLE ENERGY.

18. **ELECTROMAGNETIC SPECTRUM:**
THE COMPLETE RANGE OF FREQUENCIES OF ELECTROMAGNETIC WAVES
WHICH ARE CALLED TRANSVERSE WAVES. THEY INCLUDE RADIO,
INFRARED, VISIBLE LIGHT, ULTRAVIOLET, X-RAY, GAMMA RAYS,
AND COSMIC WAVES. THEY ARE WAVES WHICH RANGE FROM VERY
LOW FREQUENCIES TO VERY HIGH FREQUENCIES AND CAN TRAVEL
AT 3×10^8 m/s WHICH IS THE SPEED OF LIGHT IN A VACUUM.

QUESTIONS:

18-1. Which is not in the electromagnetic spectrum?
(1) visible light waves   (3) sound waves
(2) radio waves   (4) X-Rays

18-2. Which of the following electromagnetic waves has the
highest frequency?
(1) radio (2) infrared (3) X-Ray (4) visible light

18-3. As the wavelength of a visible light beam is increased from violet to red, the speed of light in a vacuum.
    (1) decreases   √(3) remains the same
    (2) increases   (4) not enough information

18-4. Gamma rays are
    (1) electrons   (3) neutrons
    (2) protons   √(4) radiation similar to X-Rays

18-5. As the color of light changes from red to blue the frequency of the light
    (1) decreases   (3) remains the same
  √(2) increases   (4) not enough information

18-6. As the frequency of an electromagnetic wave increases, its speed in a vacuum
    (1) decreases   (3) increases then decreases
    (2) increases   √(4) remains the same

18-7. All electromagnetic waves in a vacuum have the same
    (1) frequency   √(3) speed
    (2) wavelength   (4) energy

19.   **ENERGY**:
THE ABILITY OF AN OBJECT OR PERSON TO DO WORK.

QUESTIONS:

19-1. Energy is measured in the same units as
    (1) force   (2) momentum   √(3) work   (4) power
          N   *mv*   *J*   *Watt or HP*

19-2. A 500kg box and a 1000kg box are each lifted to the same level in a warehouse. Raising the heavier box requires
    (1) less energy   √(3) twice as much energy
    (2) four times as much energy (4) the same amount of energy

19-3. Which quantity and units are correctly paired?
    (1) velocity - m/s^2 (3) momentum - kg·m/s^2
  √(2) energy - kg·m^2/s^2 (4) work - N/m

$$energy - kg\, m^2/s^2$$

$$kg\, m \cdot \frac{m}{s^2}$$

$$kg\, \frac{m}{s^2} \cdot m$$

$$N \cdot m$$

19-4. Energy is
   (1) a form of power
   ✓(2) conserved in a closed system
   (3) equal to mass times velocity
   (4) measured in newtons

20.   **EXCITATION**:
   THE PROCESS OF BRINGING AN ELECTRON IN AN ATOM TO A HIGHER ENERGY LEVEL FROM A LOWER ENERGY LEVEL OR GROUND LEVEL.

   QUESTION:
20-1. Light is emitted when an electron
   (1) is boosted to a higher energy level
   ✓(2) makes a transition to a lower energy level
   (3) both (1) and (2)
   (4) neither (1) nor (2)

21.   **EXPONENTIAL GROWTH**:
   GROWTH WHICH INCREASES IN THE FORM OF A POWER OF A NUMBER.

22.   **FAHRENHEIT**:
   A TEMPERATURE SCALE IN WHICH WATER FREEZES AT 32° AND BOILS AT 212°.
   GABRIEL DANIEL FAHRENHEIT WAS A GERMAN PHYSICIST.

23.   **FIRST LAW OF THERMODYNAMICS**:
   A LAW ABOUT THE TRANSFORMATION OF HEAT INTO OTHER FORMS OF ENERGY WHICH STATES THAT IN THE TRANSFORMATION THE TOTAL AMOUNT OF ENERGY GIVEN OFF BY A SYSTEM CANNOT EXCEED THE TOTAL ENERGY ADDED TO THE SYSTEM.

   QUESTIONS:
23-1. According to the First Law of Thermodynamics, the total amount of _____ in a closed system does not change.
   (1) density   (2) buoyancy   (3) velocity   ✓(4) energy

23-2. The law of conservation of energy is expressed by the
   ✓(1) First Law of Thermodynamics (3) First Law of Motion
   (2) Second Law of Thermodynamics (4) Second Law of Motion

24. **FLUORESCENCE:**
   **THE PROCESS IN WHICH LIGHT IS GIVEN OFF BY A SUBSTANCE
   WHILE BEING ACTED UPON BY RADIANT ENERGY SUCH AS
   ULTRAVIOLET WAVES. FOR EXAMPLE, MERCURY VAPOR INSIDE
   A FLUORESCENT LIGHT IS EXCITED BY ELECTRONS BOUNCING OFF
   AN ELECTRODE IN THE TUBE. WHEN THE MERCURY VAPOR ATOMS
   DE-EXCITE THEY GIVE OFF ULTRAVIOLET WAVES. THE ELECTRONS
   FROM THIS RADIANT WAVE ENERGY CAUSE THE ATOMS IN THE
   PHOSPHOR COATING INSIDE THE FLUORESCENT TUBE TO EMIT
   RADIANT WAVE ENERGY OF A LOWER FREQUENCY. THE RESULT
   IS VISIBLE LIGHT. THEREFORE, LIGHT IS EMITTED WHEN A
   PHOSPHOR SUBSTANCE ABSORBS ELECTROMAGNETIC ENERGY AND
   THEN RE-EMITS IT AS VISIBLE LIGHT.**

   QUESTION:
24-1. When a fluorescent light is on, the mercury vapor inside
   is actually in a
   (1) liquid state   √(3) plasma state
   (2) solid state   (4) none of these

25. **FREQUENCY:**
   **THE NUMBER OF CYCLES OF A GIVEN VIBRATION OR WAVE MOTION
   WITH RESPECT TO A GIVEN AMOUNT OF TIME.**

26. **HEAT:**
   **A FORM OF KINETIC ENERGY OF MOLECULES WHICH IS REFERRED
   TO AS THERMAL ENERGY. THIS THERMAL ENERGY TRAVELS FROM
   A REGION OF HIGHER TEMPERATURE TO A REGION OF LOWER
   TEMPERATURE.**

   QUESTIONS:
26-1. Heat travels from one object to another when the objects
   have different
   (1) positrons   √ (3) temperatures
   (2) specific heats   (4) heat capacities

26-2. Heat will always flow from one object to a second object
   if the second object has a lower_____.
   (1) mass   (2) weight   (3) specific heat   (4) temperature
                                                        √

27. **INCANDESCENCE**:
GLOWING LIGHT CAUSED BY INTENSE HEAT. LIGHT IS PRODUCED
BY THE PROCESS OF EXCITATION AND DE-EXCITATION IN WHICH
RADIANT WAVE ENERGY IS EMITTED. THE FREQUENCIES OF THESE
WAVES CONTAIN THE VARIOUS FREQUENCIES OF VISIBLE LIGHT.

28. **INCOHERENT LIGHT**:
LIGHT PRODUCED BY WAVES SUCH AS THOSE FROM AN INCANDESCENT
LIGHT. THESE WAVES HAVE RANDOM ORIENTATIONS.

29. **JOULE**:
THE UNIT OF WORK OR ENERGY IN THE S.I. SYSTEM. IT IS
THE AMOUNT OF WORK DONE BY A FORCE OF ONE NEWTON ACTING
THROUGH A DISTANCE OF ONE METER IN THE DIRECTION OF THE
FORCE.

$$1J = 1N \cdot m$$

JAMES PRESCOTT JOULE WAS AN ENGLISH PHYSICIST.

QUESTION:
29-1. A unit for kinetic energy is the
  (1) Watt   (2) Joule   (3) Newton   (4) kg·m/s

30. **KELVIN SCALE**:
A TEMPERATURE SCALE IN WHICH THE ZERO POINT IS THAT OF
ABSOLUTE ZERO. EACH DEGREE IS CALLED A KELVIN.
LORD KELVIN WAS A BRITISH PHYSICIST AND MATHEMATICIAN.

QUESTION:
30-1. At which temperature is the internal energy of a substance
  at a minimum?
  (1) 0 Kelvin (2) 0°Celsius (3) 273 Kelvin (4) 0°Fahrenheit

31. **KINETIC ENERGY**:
THE ENERGY ASSOCIATED WITH A MOVING OBJECT. IT IS EQUAL
TO ONE HALF THE PRODUCT OF ITS MASS AND THE SQUARE OF
ITS VELOCITY.

$$K.E. = \tfrac{1}{2}mv^2$$

QUESTIONS:

31-1. A car is traveling at 20km/hr and another car with half the mass is traveling at 40km/hr. Which car has the greater kinetic energy?
  √(1) the lighter 40km/hr car   (3) both the same
    (2) the heavier 20km/hr car   (4) not enough information

31-2. Two of the same type cars are driven into a bale of hay. One car has twice the speed of the other car. The faster car will plow into the bale of hay
    (1) as far as the slower car   (3) twice as far
    (2) more than four times as far (4) four times as far
                                        √

31-3. Two of the same type cars are driven into a bale of hay. One car has twice the kinetic energy as the other car. The faster car will plow into the bale of hay
    (1) as far as the slower car   √(3) twice as far
    (2) more than four times as far (4) four times as far

31-4. If an object has kinetic energy, then it must be
  √(1) moving   (3) at an elevated position
    (2) at rest   (4) none of these

31-5. Which cart below has the greatest kinetic energy?
    v=5m/s   v=4m/s   v=2m/s   v=1m/s

    [1kg]   [2kg]   [4kg]   [5kg]

      A   B   C   D
    (1) A   (2) B   (3) C   (4) D
                      √

31-6. A cart of mass M traveling at speed v has kinetic energy. If the mass of the cart is cut in half and its speed is doubled, the kinetic energy of the cart will be
    (1) the same   (3) one-fourth as great
  √(2) twice as great   (4) four times as great

32.   **LASER:**
       A DEVICE THAT PRODUCES A NARROW AND VERY INTENSE BEAM OF COHERENT MONOCHROMATIC LIGHT THROUGH A PROCESS CALLED LIGHT AMPLIFICATION BY STIMULATED EMISSION OF RADIATION. **THE WORD "LASER" IS AN ACRONYM.**

32-1. A practical use for lasers is
   (1) office lights   √(3) bar-code scanning systems
   (2) household lights   (4) none of these

33. **LONGITUDINAL WAVE:**
   A TYPE OF WAVE IN WHICH THE VIBRATION OF THE MOLECULES IS PARALLEL TO THE DIRECTION OF THE WAVE MOTION. IT IS A BACK AND FORTH VIBRATION WITH THE ENERGY OF THE WAVE MOVING PARALLEL TO THAT VIBRATION.

QUESTIONS:

33-1. Longitudinal waves are involved in the transmission of
   (1) light   (2) radar   √(3) sound   (4) photons

33-2. As a longitudinal wave passes through a medium, the particles of the medium move
   (1) in circles
   √(2) parallel to the direction in which the wave travels
   (3) perpendicular to the direction of the wave
   (4) not enough information

33-3. A common property of sound waves and light waves is that
   (1) they both travel in a vacuum
   (2) they are both longitudinal waves
   (3) they are both transverse waves
   √(4) they both transfer energy

33-4. Waves in which the vibrations occur parallel to the direction of motion of the wave are called
   (1) electromagnetic waves   (3) transverse waves
   √(2) longitudinal waves   (4) radio waves

33-5. Which waves require a medium for transmission?
   (1) light   (2) infrared   (3) radio   (4) sound √

34. **PHOSPHORESCENCE:**
   THE TIME INTERVAL BETWEEN EXCITATION AND DE-EXCITATION IS DELAYED. THE ELECTRONS OF A PHOSPHORESCENT MATERIAL GET EXCITED BY EXPOSURE TO RADIANT ENERGY, SUCH AS LIGHT, AND DE-EXCITE LATER AFTER THE SOURCE OF ENERGY IS REMOVED.

QUESTION:
34-1. Some dials on clocks glow in the dark after the lights are turned off. This is known as
√(1) a time delay between excitation and de-excitation
(2) fluorescence   (3) incandescence   (4) convection

35.   **PHOTON**:
A PACKET OR BUNDLE OF RADIANT WAVE ENERGY. IT SEEMS TO EXHIBIT PARTICLE OR WAVE BEHAVIOR. RADIANT WAVE ENERGY IS CARRIED BY PHOTONS WITH EACH PHOTON CARRYING A QUANTUM OF ENERGY.

QUESTIONS:
35-1. As the wavelength of a ray of light increases, the momentum of the photons of the light ray will
√(1) decrease   (3) remain the same
(2) increase   (4) not enough information

35-2. Which color light has photons of the greatest energy?
(1) red   (2) orange   (3) green   √(4) violet

35-3. When an electron makes a transition from a higher energy level to a lower energy level, it emits _____.
(1) a photon   (2) a neutron   (3) a proton   (4) an atom
√

36.   **PLASMA**:
THE STATE OF MATTER IN WHICH ATOMS LOSE THEIR ELECTRONS AND THEREFORE ARE ELECTRICALLY CHARGED. THIS STATE REQUIRES VERY HIGH TEMPERATURES AND IS REFERRED TO AS THE FOURTH STATE OF MATTER, WITH SOLIDS, LIQUIDS AND GASES BEING THE OTHER THREE STATES OF MATTER.

37.   **POTENTIAL ENERGY**:
ENERGY OF POSITION. IT IS THE ENERGY AN OBJECT POSSESS DUE TO ITS POSITION. GRAVITATIONAL POTENTIAL ENERGY IS THE POTENTIAL ENERGY AN OBJECT POSSESS DUE TO ITS RELATIVE HEIGHT WITH RESPECT TO SOME REFERENCE LEVEL.

QUESTIONS:
37-1. As an object moves up an inclined plane, the potential energy.
(1) decreases   (3) remains the same
√(2) increases   (4) not enough information

37-2. As an object slides across a horizontal surface, the gravitational potential energy of the object will
   ✓(1) not change   (3) become greater
    (2) become less   (4) not enough information

37-3. A 10kg mass is pushed along a horizontal frictionless surface by a 20.0N force which is parallel to the surface. How much gravitational potential energy would be gained by the mass if it is moved one meter horizontally?
   ✓(1) 0 J   (2) 4 J   (3) 6 J   (4) 40 J

38. **POWER:**

THE RATE OF DOING WORK. THE UNIT OF POWER IS THE WATT.
POWER = WORK/TIME
WATT = JOULE/SECOND
JAMES WATT WAS A SCOTTISH ENGINEER AND INVENTOR.

QUESTIONS:

38-1. If two identical jobs are done in different amounts of time, they would require the same amount of work but they would require different amounts of
   (1) energy   (2) momentum   (3) power   (4) none of these
                                 ✓

38-2. As the time required for a person to run up a flight of stairs decreases the power developed by this person
   (1) decreases   (3) remains the same
  ✓(2) increases   (4) none of these

38-3. The rate at which work is done is called
   (1) momentum   (3) kinetic energy
   (2) potential energy   ✓(4) none of these

38-4. The rate of doing work may be measured in
  ✓(1) Watts   (2) Newtons   (3)Joules   (4) kilocalories

39. **POWER PRODUCTION:**

   FOSSIL FUEL:

     THE REMAINS OF PLANTS AND ANIMALS THAT LIVED MILLIONS OF YEARS AGO. THESES UNDERGROUND DEPOSITS ARE COAL, CRUDE OIL, AND NATURAL GAS.

       COAL:

     A ROCK THAT MAKES A LOT OF HEAT WHEN BURNED. IT

IS DRILLED FROM MINES DEEP BELOW THE EARTH'S SURFACE.
AIR POLLUTION AND ACID RAIN ARE NEGATIVE FEATURES
WHEN USING COAL AS A FUEL.

CRUDE OIL:

OBTAINED FROM WELLS. OIL REFINERIES TURN THE CRUDE
OIL INTO GASOLINE AND OTHER PRODUCTS. IT, TOO, CAUSES
AIR POLLUTION.

NATURAL GAS:

ALSO KNOWN AS METHANE. PUMPED FROM WELLS, NATURAL
GAS IS USUALLY FOUND IN ASSOCIATION WITH CRUDE OIL
DEPOSITS. IT CAN ALSO BE PRODUCED BY BACTERIAL
DECAY OF VEGETATION AND ANIMAL WASTES IN THE ABSENCE
OF AIR. THERE IS NO AIR POLLUTION WHEN IT IS BURNED
AS A FUEL. NATURAL GAS IS THEREFORE CALLED A CLEAN
FUEL AND IS USED FOR COOKING.

ETHANOL - ALSO CALLED GRAIN ALCOHOL:

OBTAINED FROM THE FERMENTATION OF GRAINS SUCH AS CORN.

GASOHOL:

A MIXTURE OF ETHANOL AND GASOLINE.

GEOTHERMAL POWER:

OBTAINED BY DRILLING HOLES IN THE EARTH'S CRUST.
THERE IS LOW HEAT CONDUCTIVITY OF THE ROCKS. IT IS
RELATIVELY POLLUTION FREE.

HYDROELECTRIC POWER:

THE RUSH OF A WATERFALL. THERE ARE FEW GOOD LOCATIONS
FOR A HYDROELECTRIC POWER PLANT AND CONSTRUCTION COSTS
ARE HIGH. ALTHOUGH THERE IS RELATIVELY NO AIR
POLLUTION, IT HAS A NEGATIVE IMPACT ON MARINE AND
PLANT LIFE IN THE WATER.

SOLAR POWER:

RADIATION FROM THE SUN. PHOTOELECTRIC CELLS CONVERT
THE SUN'S RAYS INTO ELECTRICITY. SOLAR ENERGY MUST
BE STORED BY HEATING WATER, ROCKS OR GRAVEL, ETC.
SOLAR ENERGY COULD BE VERY EXPENSIVE.

NUCLEAR FISSION:

THE SPLITTING OF THE NUCLEUS OF A PLUTONIUM OR
URANIUM ATOM WITH THE RELEASE OF HUGE AMOUNTS OF
ENERGY.

NUCLEAR REACTOR:

METHOD OF RELEASING ENERGY AT A CONSTANT RATE USING
CONTROLLING RODS. ALTHOUGH IT DOES NOT ADD POLLUTANTS
TO THE AIR, IT MAY CAUSE THE HEATING OF WATER IN LOCAL
RIVERS WHICH WOULD AFFECT PLANT AND MARINE LIFE. IN

ADDITION, RADIOACTIVE WASTES MUST BE STORED UNTIL
THEY ARE NO LONGER RADIOACTIVE.

BREEDER REACTOR:

NUCLEAR REACTOR WHICH GENERATES NUCLEAR ENERGY WHILE
IT ALSO PRODUCES MORE FISSIONABLE MATERIAL.

NUCLEAR FUSION:

TWO NUCLEI ARE FUSED TO FORM ONE NUCLEUS. THE ENERGY
OF THE SUN IS RELEASED BY FUSION GIVING LIGHT AND
RADIANT WAVE ENERGY. EXTREMELY HIGH TEMPERATURES
ARE REQUIRED FOR THE HUGE AMOUNT OF ENERGY THAT IS
RELEASED. NO RADIOACTIVE WASTES ARE PRODUCED BY
THE FUSION PROCESS.

THERMONUCLEAR POWER:

POWER RESULTING FROM FUSION. NO AIR POLLUTION AND
NO RADIOACTIVE WASTES ARE PRODUCED.

QUESTIONS:

39-1. Fossil fuels most likely were formed over millions of
years ago from
   (1) rocks found on mountain tops
   (2) glaciers
   (3) clay
 √ (4) remains of dead plants and animals

39-2. _____ is a clean fuel.
   (1) Coal   (2) Oil   (3) Natural Gas   (4) none of these
                           √

39-3. The energy source that causes the molecules of the
atmosphere to stay in motion is
   (1) air pressure   (3) their own kinetic energy
 √ (2) solar power   (4) none of these

39-4. A series of rapid nuclear fissions is called
 √ (1) a chain reaction   (3) nuclear fusion
   (2) decay   (4) radioactive wastes

39-5. Radioactive by-products are the result of atomic nuclei
 √ (1) breaking apart   (3) neither (1) nor (2)
   (2) combining   (4) not enough information

39-6. The energy released by the Sun is the result of atomic nuclei
   (1) breaking apart   (3) neither (1) nor (2)
 ✓ (2) combining   (4) not enough information

39-7. The fission process in a nuclear reactor is controlled by regulating the number of available
   (1) electrons   (2) neutrons   (3) protons   (4) positrons
                      ✓

39-8. In a nuclear reactor, the function of the controlling rod is to
   (1) slow down neutrons   (3) produce neutrons
   (2) absorb neutrons   (4) speed up neutrons
      ✓

40. **RADIATION:**
THE FLOW OF ENERGY IN THE FORM OF RAYS OR WAVES. IN A IN A VACUUM THESE WAVES TRAVEL AT THE RATE OF 3×10^8 m/s.

QUESTION:
40-1. Heat from the Sun gets to the Earth by the process of
   (1) conduction   (2) convection   (3) radiation   (4) osmosis
                                          ✓

41. **SECOND LAW OF THERMODYNAMICS:**
THERMODYNAMICS IS THE TRANSFORMATION OF THERMAL ENERGY INTO MECHANICAL ENERGY. THE SECOND LAW STATES THAT THE TOTAL AMOUNT OF WORK OBTAINED FROM THIS CHANGE IS LESS THAN THE TOTAL AMOUNT OF ENERGY USED. IN OTHER WORDS, SOME OF THE ENERGY BECOMES LOW LEVEL THERMAL ENERGY WHICH CANNOT DO WORK. NOTHING IS 100% EFFICIENT.

QUESTION:
41-1. It is _____ possible to completely convert a given amount of thermal energy into mechanical energy.
   (1) sometimes   (2) always   (3) never   (4) none of these
                                   ✓

42. **TEMPERATURE:**
A MEASURE OF THE INTERNAL KINETIC ENERGY OF AN OBJECT. IT IS MEASURED USING A THERMOMETER WITH EITHER THE CELSIUS, FAHRENHEIT, OR KELVIN SCALES. THE FOLLOWING ARE CONVERSION METHODS FOR THE CELSIUS AND THE FAHRENHEIT SCALES.

TO CONVERT FAHRENHEIT TO CELSIUS, USE THE FOLLOWING FORMULA: $C = \frac{5}{9}(F - 32)$

TO CONVERT CELSIUS TO FAHRENHEIT, USE THE FOLLOWING FORMULA: $F = \frac{9C}{5} + 32$

QUESTIONS:

42-1. The numerical readings of Celsius and Fahrenheit thermometers are the same at
   (1) 4°   (2) -40°   (3) 40°   (4) 0°

42-2. A temperature of 77°F is equivalent to
   (1) 20°C   (2) 25°C   (3) 30°C   (4) 35°C

42-3. A temperature of 37°C is equivalent to
   (1) 98.6°C   (2) 96.8°F   (3) 98.2°F   (4) 98.6°F

43.   **TRANSVERSE WAVES:**
      A WAVE IN WHICH THE PARTICLES OF MATTER VIBRATE AT RIGHT ANGLES TO THE MOTION OF THE WAVE.

QUESTIONS:

43-1. When a transverse wave is moving through a medium, how do the particles move?
   (1) They vibrate parallel to the direction of the wave.
   (2) They vibrate perpendicular to the wave.
   (3) They do not move.
   (4) None of theses.

43-2. 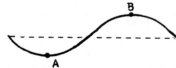   As the wave travels, what is transferred between point A and point B?
   (1) mass and energy   (3) mass
   (2) energy   (4) nothing

44.   **WAVELENGTH:**
      THE DISTANCE BETWEEN TWO SUCCESSIVE CRESTS OR TROUGHS OF THE SAME WAVE WHERE THE CREST IS THE HIGHEST POINT OF THE WAVE AND THE TROUGH IS THE LOWEST POINT OF THE WAVE.

QUESTION:

44-1. In the diagram at the right, wave A has the same wavelength as wave _____.
(1) B (2) C (3) D (4) not enough information

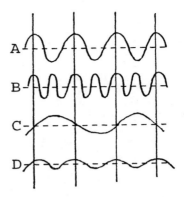

45. **WORK:**

THE PRODUCT OF THE FORCE EXERTED ON AN OBJECT MULTIPLIED BY THE DISTANCE THE OBJECT MOVED IN THE DIRECTION OF THE FORCE.

WORK = FORCE x DISTANCE

WORK IS MEASURED IN JOULES.

QUESTIONS:

45-1. Lifting a 40kg sack a vertical distance of 2m requires _____ lifting a 20kg sack a vertical distance of 4m?
(1) the same work as
(2) more work than
(3) less work than
(4) not enough information

45-2. A 2.5kg mass is pulled on a frictionless surface as shown in the diagram at the right by a 20N force through a distance of 5.0m. What amount of work is done?
(1) 5.0J   (2) 12.5J   (3) 100J   (4) 50J

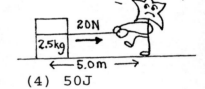

45-3. You push against a rock with a force of 50N for 30 seconds. If the rock does not move, how much work have you done?
(1) 1500J
(2) 5/3J
(3) 3/5J
(4) zero J

45-4. The same units can be used to measure work and
(1) power   (2) force   (3) energy   (4) weight

1. Which example would you estimate to have the greatest sum, Example 1 or Example 2?

Example 1	**Example 2**
987654321	123456789
87654321	12345678
7654321	1234567
654321	123456
54321	12345
4321	1234
321	123
21	12
1	1
_____	_____

2. Place a symbol between 7 and 8 to get a number greater than 7 and less than 8.

3. How much dirt is there in a hole 2ft by 2ft by 2ft?

4. What day would it be if Thursday was 4 days before the day after tomorrow?

5. A student came into class to take a hundred question true-false test and saw that he did not know any of the answers. He decided to toss a coin to get his answers: heads = true, tails = false. When he finished the exam, the teacher saw him put the coin away and she thought he would turn in his paper. But, instead, he reached into another pocket and took out a different coin. He then proceeded to toss the new coin for heads and tails again. The teacher turned to him and said, "What are you doing now?" He looked up and said, "I'm checking my answers."

Chapter XV

PRACTICE FINAL EXAM QUESTIONS

1. If an object is lifted 15 meters, it has potential energy. If it is lifted 30 meters it has
    - (1) as much energy
    - (2) less energy
    - (3) four times as much energy
    - (4) twice as much energy ✓

2. The gravitational potential energy
    - (1) is independent of height
    - (2) is always positive
    - (3) is independent of path ✓
    - (4) decreases with increasing height

3. Energy is
    - (1) a tangible thing
    - (2) a vector quantity
    - (3) an ability to do work ✓
    - (4) not associated with fundamental forces

4. Work requires
    - (1) a zero net force
    - (2) motion ✓
    - (3) zero momentum
    - (4) a force perpendicular to the direction of motion

5. The unit of work is
    - (1) the pound
    - (2) the joule ✓
    - (3) the newton
    - (4) meter/sec

6. Solar power is actually
    - (1) wind power
    - (2) fission power
    - (3) fusion power ✓
    - (4) electric power

7. Suppose the doubling time for the growth of bacteria in a jar is one minute and suppose the jar starts with one bacterium and fills to capacity in 60 minutes. How many minutes are required for the jar to fill to capacity if the jar starts with four bacterium?
    - (1) 58 minutes ✓
    - (2) 56 minutes
    - (3) 30 minutes
    - (4) 15 minutes

8. If your wages are one dollar for the first day, two dollars for the second day, four dollars for the third day and so on, doubling for a total of eight days. Your combined wages for the eight days of work will be
    - (1) $255 ✓
    - (2) $128
    - (3) $127
    - (4) $256

$$2^8 - 1 = 255$$

9. On a windy day compared to a non-windy day atmospheric pressure is
   (1) less   (2) unchanged   (3) greater   (4) faster
10. An open umbrella held straight and not tilted on a windy day has a tendency to move upward mainly because
    (1) the umbrella is pushed upward by the wind
    (2) moving air over the top of the curved surface causes a reduced downward pressure against the top
    (3) air moves up into the bottom of the umbrella and increases the pressure there
    (4) the umbrella can turn inside out
11. The kinetic energy of a moving mass is proportional to the
    (1) velocity   (3) square root of the velocity
    (2) square of the velocity   (4) velocity halved
12. The rising of warm air has to do with the fact that faster moving molecules tend to move in regions of less
    (1) pressure (2) opposition (3) density (4) all of these
13. The source of all wave motion is a
    (1) harmonic body   (3) variation
    (2) wave pattern   (4) vibrating body
14. Some light switches glow in the dark after the lights are turned off because of the effect of
    (1) fluorescence   (3) a time delay between excitation and de-excitation
    (2) incandescence   (4) lasers
15. Work is equal to
    (1) force X time   (3) energy X time
    (2) power X time   (4) force X mass
16. One person can do a certain amount of work in a given time. A second person does the same work in half that time. The power developed by the second person relative to the first is _____as much.
    (1) one-half   (3) four times
    (2) one-fourth   (4) twice
17. The work done by a force acting on an object is defined as the product of the force and the distance through which the object moves _____to the force.
    (1) perpendicular   (3) opposite
    (2) parallel   (4) in any direction
18. As the temperature of a substance increases, the average kinetic energy of the molecules of the substance
    (1) decreases   (3) remains the same
    (2) increases   (4) disappears

19. If an atom of uranium splits into two parts, it is an example of
    (1) alpha decay   ✓(3) fission
    (2) beta decay   (4) fusion
20. Heat transfer takes place because of a difference in
    (1) potential energy   (3) specific heat
    ✓(2) temperature   (4) heat content
21. Referring to the diagram at the left, the sum of the kinetic and potential energies of the bob at position 1 is _____ the kinetic and potential energies of the bob at position 2.
    ✓(1) the same as   (3) greater than
    (2) less than   (4) not enough information
22. The reduced air pressure above the wing of a moving airplane can be explained by a principle introduced by
    (1) Torricelli (2) Einstein (3) Bernoulli (4) Archimedes
23. The process whereby certain materials absorb ultraviolet waves and emit waves with a lower frequency corresponding to the visible portion of the electromagnetic spectrum is known as
    (1) polarization   ✓(3) fluorescence
    (2) photon   (4) phosphorescence
24. The principle on which the hydrogen bomb is based is
    (1) fission   (3) magnetism
    ✓(2) fusion   (4) Bernoulli's principle
25. The (N·m)/sec is a unit of
    (1) work   ✓(3) power
    (2) energy   (4) efficiency

MATH HUMOR XV

1. In the following diagram, find the perimeter of the small squares if the perimeter of the large square is one.

2. If 1+2+3-4+5+6+78+9=100, find another arrangement of the nine digits in the same order which equals 100.

3. Using the diagram at the right, find AB if the radius of the circle is five.

AB is the diagonal of the square in the diagram.

4. The number 720 has 30 different divisors. Can you find them?

5. The number 840 has 32 different divisors. Can you find them?

6. How much is three plus three multiplied by three?

7. Compute the answers as quickly as you can.

```
        1X9 + 2 =
       12X9 + 3 =
      123X9 + 4 =
     1234X9 + 5 =
    12345X9 + 6 =
   123456X9 + 7 =
  1234567X9 + 8 =
 12345678X9 + 9 =
123456789X9 + 9 =
```

Answers

CHAPTER I

EXAMPLES FOR THOUGHT I

HELP !!!

2. (1)
3. 90'
4. (2)
5. (4)
7. $(3.5m)^2 + (5.8m)^2 = s^2$; $s = 6.8m$
    $$\tan \theta = \frac{5.8m}{3.5m}$$
    $= 59°$;   Direction = S31°E

MATH HUMOR I

1. XII -- = VII
2. IV - I = V

    OR

4 angles {take away} 1 angle equals 5 angles

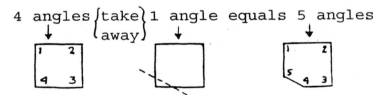

3. The most common answer is six. However, the answer could
    be five if the cut is made along the diagonal. Actually
    there could be many answers if steps are cut along the
    diagonal.

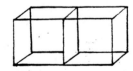

Answer = 6
Vertical cut

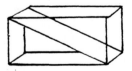

Answer = 5
Cut along the
diagonal

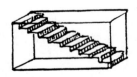

Steps along the
diagonal

CHAPTER II

EXAMPLES FOR THOUGHT II

1. (1)
2. (4)
3. (a) $d = \frac{1}{2}at^2$; $d = \frac{1}{2}(7m/s^2)(5s)^2$
   $= (3.5m/s^2)(25s^2)$
   $= 87.5m$

   (b) $v = at$; $v = (7m/s^2)(5s)$
   $= 35m/s$

4. (a) $d = \frac{1}{2}gt^2$; $d = \frac{1}{2}(9.8m/s^2)(3.5s)^2$
   $= (4.9m/s^2)(12.25s^2)$
   $= 60.025m$
   $= $ approx. $60m$

   (b) $v = gt$; $v = (9.8m/s^2)(3.5s)$
   $= 34.30m/s$
   $= $ approx. $34m/s$

5. (a) $d = \frac{1}{2}gt^2$; $d = \frac{1}{2}(9.8m/s^2)(10s)^2$
   $= (4.9m/s^2)(100s^2)$
   $= 490m$

   (b) $v = gt$; $v = (9.8m/s^2)(10s)$
   $= 98m/s$

   (c) In the 10 seconds the object falls 490m.
   In 9 seconds it would fall
   $d = \frac{1}{2}(9.8m/s^2)(9s)^2$
   $= (4.9m/s^2)(81s^2)$
   $= 396.9m$

   Thus, 490m - 396.9m = 93.1m, which is the amount it would fall in the last second.

6. A car that is moving in a circular path.
7. (3)

MATH HUMOR II

1. Ans 1: $101 - 10^2 = 1$ (Move the two to the exponent position.)
   Ans 2: $101 = 102 - 1$ (Move the bottom of the equal sign.)
2. SEVEN - Take off the "S" leaves "EVEN".
3. The teacher said "John is very smart".
   "The teacher," said John, "is very smart".
   Without punctuation marks, the sentence is ambiguous.
4. The third "that".
5. "S". Read each square clockwise. Together they say, "This puzzle is made of square"s".

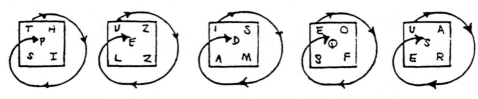

CHAPTER III

EXAMPLES FOR THOUGHT III

1. (2)
2. (2)
3. (3)
4. (3)
5. (3)
6. (4)
7. Theoretically, you should land on your footprints because your forward motion should remain constant and be the same as that of the plane. It would seem as though you jumped vertically up and came vertically down on the same spot. However, to someone observing you jump from outside the plane, your body moved in a path called a parabola. There was an upward and forward motion followed by a downward and forward motion.

MATH HUMOR III

1. The left side is twenty and the right side is twenty-two or twenty-(too).
2. 10T010 = 9:50 (Ten to Ten equals 9:50)
3. Place "Z" above the line since all letters above the line are formed with straight lines and below the line they are formed with curved lines.

CHAPTER IV

EXAMPLES FOR THOUGHT IV

1. tangent to the surface
2. (3)
3. (2)
4. (2)

5. $F = \dfrac{Gm_1 X m_2}{d^2}$

$\quad = \dfrac{6.67 \times 10^{-11} N \cdot m^2}{kg^2} \cdot \dfrac{(4800kg)(21000kg)}{(4.0 \times 10m)^2}$

$\quad = 4.2 \times 10^{-6} N$

6. $F = \dfrac{Gm_1 X m_2}{d^2}$

$\quad = \dfrac{6.67 \times 10^{-11} N \cdot m^2}{kg^2} \cdot \dfrac{(50kg)(40kg)}{(4m)^2}$

$\quad = 8.3 \times 10^{-9} N$

This is insufficient to pull two insects towards each other.

MATH HUMOR IV

2. ZERO, because one of the factors will be (x - x).
3. Twelve. (Jan 2nd, Feb. 2nd, Mar. 2nd, etc.)
4. The rule that the square root of a quotient is equal to the quotient of the square roots does not apply when dealing with imaginary numbers.

CHAPTER V

EXAMPLES FOR THOUGHT V

1. Light travels at about 186,000 miles per second or 3×10^8 meters per second, roughly about seven times around the world in one second. Therefore, we could hit the Moon in a little over 1¼ seconds and the Sun in a little over 8 minutes.

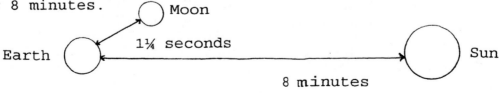

2. In extending your arms, you increase your rotational inertia by distributing more mass farther from the axis of rotation (along the rail). When you start to rotate the increased

inertia resists the change and gives you more time to regain your balance. On a narrow fence, gravity would take over and you would begin to fall downward.

3. The first motion is the horizontal movement of the bullet with a velocity of 100m/s. The second motion is vertically downward because of the pull of gravity which accelerates at the rate of $9.8m/s^2$. It will fall to the ground in exactly 4 seconds ignoring air resistance.

Horizontal: Distance = vt
Vertical: Distance = $\frac{1}{2}gt^2$

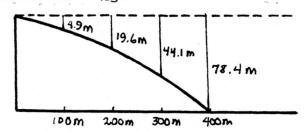

DOWN	ACROSS
after 1 sec. fall = 4.9m	after 1 sec. it moves 100m
after 2 sec. fall = 19.6m	after 2 sec. it moves 200m
after 3 sec. fall = 44.1m	after 3 sec. it moves 300m
after 4 sec. fall = 78.4m	after 4 sec. it moves 400m

4.

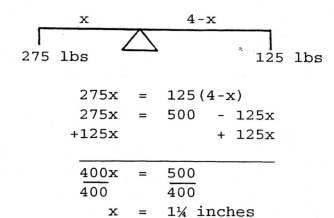

$$275x = 125(4-x)$$
$$275x = 500 - 125x$$
$$+125x \qquad + 125x$$

$$\frac{400x}{400} = \frac{500}{400}$$
$$x = 1\frac{1}{4} \text{ inches}$$

6. (2)
7. (3)
8. (3)

MATH HUMOR V

1. A "V". GRA"V"ITY
2. It takes no time at all. The inch worm is already at the last page of volume II when he starts. If the books are arranged in numeric order from left to right on the shelf, then the first page of volume I is up against the last page of volume II.

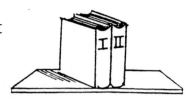

3. Because seven "eight" nine (seven ate nine).
4. It takes 16 minutes. If it takes 12 minutes to cut four parts it means only three cuts are made and each cut takes four minutes. To cut a board into five parts requires four cuts and at the rate of four minutes a cut the answer is 16 minutes.

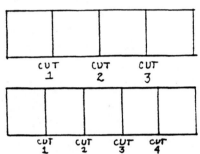

5. The last digit of the answer could only come out to 1, 2, 3, or 4 which corresponds to the 6, 7, 8, and 9 for clubs, hearts, spades and diamonds, respectively, in the instructions. The other portion of the number is one more than the card selected. For example, a 53 would be a four of spades.

CHAPTER VI

EXAMPLES FOR THOUGHT VI

1. (3)
2. (1)
3. (4)

4.

$$\leftarrow \quad 4.0kg \qquad \qquad (-) \qquad \qquad (+) \qquad 0.05kg \quad \rightarrow$$

$m_{(bullet)} \quad = 0.05kg$

$m_{(gun)} \quad = 4.0kg$

$v_{(bullet)} \quad = 800m/s$

$v_{(gun)} \quad = v$

momentum before = momentum after

$mv_{before} = mv_{after}$

$0 = +(0.05kg)(800m/s) - (4.0kg)(v)$

$0 = +(40kg)(m/s) - (4.0kg)(v)$

$$\frac{+(4.0kg)(v)}{(4.0kg)} = \frac{+(40kg)(m/s)}{(4.0kg)}$$

$$v = 10m/s$$

5. No. Friction would be an external force so the momentum would not be conserved. The skater would gradually slow down.

6. (4)

7. momentum before = momentum after

$$0 = -(40kg)(v) + (10kg)(4m/s)$$
$$(40kg)(v) = (10kg)(4m/s)$$
$$v = 1m/s$$

8. momentum before = momentum after

$$(40kg)(5m/s) = (40+10kg)(v)$$
$$v = 4m/s$$

9. momentum before = momentum after

$$(4kg)(5m/s) = (4kg)(-4m/s) + (10kg)(v)$$
$$v = 3.6m/s$$

MATH HUMOR VI

1. Method I: Move one of the lines that make the V.
                 (V becomes X)
                 II = XI

   Method II: Take the first vertical line and place it horizontally over the last vertical line.
                 I = VT

2. 6 = 111111 and 5 = 11111. Using all the lines we could write the word

3. So that he could win the no bell (Nobel) prize.

4. $\sin x = 6n$ <u>Method I:</u>

Divide both sides of the equation by "n". (Division by "n" in this manner is not mathematically correct but it is meant as humor.)

$$\frac{\sin x}{n} = \frac{6n}{n}$$

<u>Method II:</u>
Switch the "n" and the "x" in sin x.
sin x = 6n becomes six n = 6n
(Remember, this is only a joke.)

5. 31

CHAPTER VII

MIDTERM REVIEW VII

1-1 (4)	1-2 (2)	1-3 (4)	1-4 (4)	1-5 (4)
2-1 (3)	2-2 (1)	2-3 (2)	2-4 (4)	2-5 (4)
7-1 (2)	7-2 (3)	7-3 (3)	7-4 (4)	7-5 (2)
7-6 (2)	7-7 (3)	8-1 (4)	8-2 (3)	10-1 (3)
12-1 (4)	12-2 (2)	13-1 (3)	13-2 (3)	13-3 (3)
13-4 (3)	13-5 (3)	15-1 (4)	15-2 (2)	15-3 (2)
16-1 (2)	16-2 (2)	16-3 (3)	17-1 (4)	17-2 (1)
17-3 (3)	17-4 (1)	17-5 (3)	19-1 (3)	19-2 (1)
20-1 (3)	20-2 (3)	20-3 (4)	21-1 (3)	22-1 (2)
22-2 (2)	22-3 (3)	22-4 (2)	22-5 (2)	22-6 (3)
22-7 (4)	22-8 (1)	22-9 (3)	23-1 (4)	23-2 (4)
24-1 (4)	24-2 (4)	24-3 (4)	24-4 (4)	24-5 (1)
24-6 (2)	24-7 (3)	24-8 (2)	24-9 (1)	24-10 (2)
27-1 (3)	27-2 (1)	27-3 (4)	27-4 (4)	27-5 (2)
31-1 (2)	34-1 (4)	35-1 (4)	36-1 (1)	36-2 (3)
36-3 (4)	36-4 (4)	37-1 (1)	37-2 (4)	37-3 (2)
37-4 (4)	37-5 (2)	37-6 (1)	37-7 (2)	37-8 (1)
37-9 (1)	38-1 (4)	38-2 (4)	38-3 (2)	38-4 (3)
38-5 (4)	38-6 (2)	38-7 (1)	38-8 (4)	38-9 (2)
39-1 (1)	41-1 (1)			

MATH HUMOR VII

1. "E" Each letter is the first letter of the counting numbers. One, Two, Three, etc.
2. Turn the paper upside down with the numbers on it and circle the row 1 1 1 and 6 6 6.
4. Ans: "Gee, I'm a tree." (G-E-O-M-E-T-R-Y)

CHAPTER VIII

PRACTICE MIDTERM QUESTIONS VIII

1. (3)	2. (2)	3. (4)	4. (3)
5. (1)	6. (3)	7. (3)	8. (4)
9. (3)	10. (1)	11. (4)	12. (4)
13. (1)	14. (4)	15. (4)	16. (4)
17. (4)	18. (2)	19. (2)	20. (1)
21. (4)	22. (1)	23. (2)	24. (1)
25. (3)			

MATH HUMOR VIII

1. (8888-888) divided by 8 = 1000 OR

$$\begin{array}{r} 888 \\ 88 \\ 8 \\ 8 \\ 8 \\ \hline 1000 \end{array}$$

2. VIII   ⓪ ②③①⑤ LINES \ / | | |
3. The answer is always 11 times the 7th number.
4. The next symbol is 5 attached to its mirror image. ⟶ ♂

5. 55 because the opposite sides of a die add up to seven.

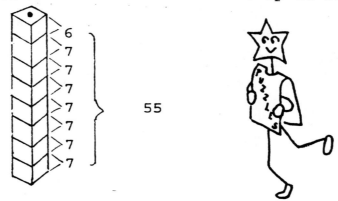

$$6$$
$$7$$
$$7$$
$$7$$
$$7$$
$$7$$
$$7$$
$$7$$

$$55$$

CHAPTER IX

EXAMPLES FOR THOUGHT IX

1. (4)
2. (1)
3. (4)
4. G.P.E. = mgh
   = $(10kg)(9.8m/s)(2m)$
   = 196J
5. K.E. = $\frac{1}{2}mv^2$
   = $\frac{1}{2}(5kg)(10.5m/s)^2$
   = 275.6J
6. (3)
7. (3)
8. (3)
9. (2)
10. The stopping distance of the faster car is four times as great.

   Car I: v_1 = 90mph; d_1 = distance to stop car I
   Car II: v_2 = 45mph; d_2 = distance to stop car II
   Applying the same force to stop "F", we get
   $Fd_1 = \frac{1}{2}m(v_1)^2 = \frac{1}{2}m(90)^2 = \frac{1}{2}m(8100)$
   $Fd_2 = \frac{1}{2}m(v_2)^2 = \frac{1}{2}m(45)^2 = \frac{1}{2}m(2025)$
   The force is the same and the distances are different with the second distance being four times greater than the first.

MATH HUMOR IX

1. ONE THOUSAND
2. The middle digit is always 9 and the first and last digit add up to 9.
3. 1089
4. 98,765
   +1,234
   99,999
5. Twenty - **9, 19, 29, 39, 49, 59, 69, 79, 89,**
   90, 91, 92, 93, 94, 95, 96, 97, 98, 99
6. 9 X 8 + 7 + 6 + 5 + 4 + 3 + 2 + 1 + 0 = 100

CHAPTER X

EXAMPLES FOR THOUGHT X

1. $F = \dfrac{9C}{5} + 32$ Subtract 32 from both side of the equation.

   $F - 32 = \dfrac{9C}{5}$ Multiply both sides of the equation by $\dfrac{5}{9}$

   $\dfrac{5}{9}(F - 32) = C$ OR $C = \dfrac{5}{9}(F - 32)$

2. In the formula for finding Celsius the 5/9 is slightly over 1/2 and multiplying by 1/2 is equivalent to dividing by 2.
   In the formula for finding Fahrenheit the 9/5 is close to 2.
3. F = 98.6°, normal body temperature
4. C = 30
5. (2)
6. (3)
7. It will expand. Suppose each of the other squares were cut out, heated and then put together in the original format. It would be obvious that the center square space would be larger.

8. Using the formula   C = (5/9)(F-32)

                    if the easy method = the actual formula
             we get   ½(F-30) = (5/9)(F-32)
          mult. by 18: 9(F-30) =   10(F-32)
     distributive law: 9F-270 =   10F-320
                              50° = F

    Using the formula   F = (9/5)C + 32

                    if the easy method = the actual formula
                 we get 2C + 30 = (9/5)C + 32
              mult. by 5: 10C + 150 =   9C + 160
                                C = 10°

    Thus the approximation method and the actual formulas
    give the same value for a Fahrenheit reading of 50° or a
    Celsius reading of 10°.

9. -40°F. This is the only point where the Celsius
    and the Fahrenheit scales give the same numeric
    value.

MATH HUMOR X

1. Summer days hot and longer
    (expanded)
    Winter days are cold and
    shorter (contracted)
2. Heat travels faster because
    you can always catch a cold.
    (Note: Remember, only heat
    travels from a region of higher
    temperature to a region of lower
    temperature.)

CHAPTER XI

EXAMPLES FOR THOUGHT XI

1. (3)

2.　　　　　DISTANCE = SPEED X TIME

$$\frac{93,000,000 \text{miles}}{186,000 \text{miles/sec.}} = \frac{(186,000 \text{miles/sec.})(time)}{186,000 \text{ miles/sec.}}$$

　　　　　　　　　500sec. = time

or　　　　　　　　time = 500sec.

Since 60sec = 1 minute, we can form the following proportion

$$\frac{60\text{sec}}{500\text{sec}} = \frac{1 \text{ minute}}{x \text{ minutes}}$$

　　　　　　　　60x = 500

then　　　　6x = 50

　　　　　　　x = approximately 8

MATH HUMOR XI

1.　The letter "n".

2.　The quick way to find 111111111^2 is to use the following pattern.

1^2	=	1
11^2	=	121
111^2	=	12321
1111^2	=	1234321
11111^2	=	123454321
111111^2	=	12345654321
1111111^2	=	1234567654321
11111111^2	=	123456787654321
111111111^2	=	12345678987654321

3.　H to O　　(H_2O)

4.　(2) because $10^0 = 1$ ($x^0 = 1$ by definition except when x=0)

　　　　and $1^{10} = 1$ (1^{10} means $1 \cdot 1 \cdot 1 \cdot 1 \cdot 1 \cdot 1 \cdot 1 \cdot 1 \cdot 1 \cdot 1$ which

　　　　　　equals 1)

CHAPTER XII

MATH HUMOR XII

1.　"Don't bother me.　I have my own problems."

3. It reads "BIG SHOES".
4. 6,3,2,0 (The numbers are in alphabetical order.)
5. The last two digits equal your age and the first two digits
   equal the month of your birthday.
   Example: 1260 means the person was 60 in December.

CHAPTER XIII

EXAMPLES FOR THOUGHT XIII

2. $A = P(1+r)^n$ P = principal
                          r = rate of interest
                          n = number of compounding periods

$A = 1.00(1+.02)^{1997}$

$= 1.49466 \times 10^{17}$

= approx. \$149,000,000,000,000,000

3. \$1.00 at 2% = \$.02
    Each year adds on \$.02
    1997 X \$.02 = 39.94
                    +1.00

                    \$40.94

4. If you took the penny, you would have about \$10 million
    after 30 days.

5.

0 FOLDS	1 FOLD	2 FOLDS	3 FOLDS	4 FOLDS	5 FOLDS
1 LAYER	2 LAYERS	4 LAYERS	8 LAYERS	16 LAYERS	32 LAYERS
(2^0)	(2^1)	(2^2)	(2^3)	(2^4)	(2^5)

After nine folds the number of layers would be 2^9 which
equals 512. This is thicker than this book. It would be
difficult to fold in half.

6. Answer: 58 seconds

Starting with one microbe
After 1 second there are 2 microbes (2^1)
" 2 " " " 4 " (2^2)

.
.
.

" 60 " " " " (2^{60})

60 seconds equals one minute at which time there are 2^{60} microbes and the jar is full.

Starting with 4 microbes
After 1 second there are 8 microbes (2^3)
" 2 " " " 16 " (2^4)
" 3 " " " 32 " (2^5)

.
.
.

" 58 " " " (2^{60})

2^{60} microbes fills the jar as explained above. Starting with 4 microbes, the 2^{60} is reached in 58 seconds.

7. The lake would be fully covered on the 60th day. Therefore, it would be half covered on the 59th day and one quarter covered on the 58th day.

MATH HUMOR XIII

1. Saw the wood in half. Two halves make a whole (joke - hole) and crawl through.

2. H_2O is for drinking. The joke is the misuse of the "four".

4. 1 HR. Ex. 7am, 7:30am, 8am

5. 43

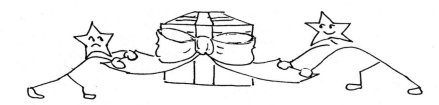